I0759582

MYRIAD, MICROSCOPIC AND MARVELLOUS

MYRIAD, MICROSCOPIC AND MARVELLOUS

The World of Antoni van Leeuwenhoek

GEERTJE DEKKERS

Translated by Andy Brown

REAKTION BOOKS

Published by
REAKTION BOOKS LTD
2–4 Sebastian Street
London EC1V 0HE, UK
www.reaktionbooks.co.uk

First published in English 2025

English-language translation by Andy Brown

Original title: Veel, klein en curieus. Translated from the Dutch language
www.spectrumboeken.nl

This book was published with the support
of the Dutch Foundation for Literature

Nederlands letterenfonds
dutch foundation
for literature

EU GPSR Authorised Representative
Logos Europe, 9 rue Nicolas Poussin, 17000, La Rochelle, France
email: contact@logoseurope.eu

Printed and bound in Great Britain by Bell & Bain, Glasgow

A catalogue record for this book is available from the British Library

ISBN 978 1 83639 097 8

CONTENTS

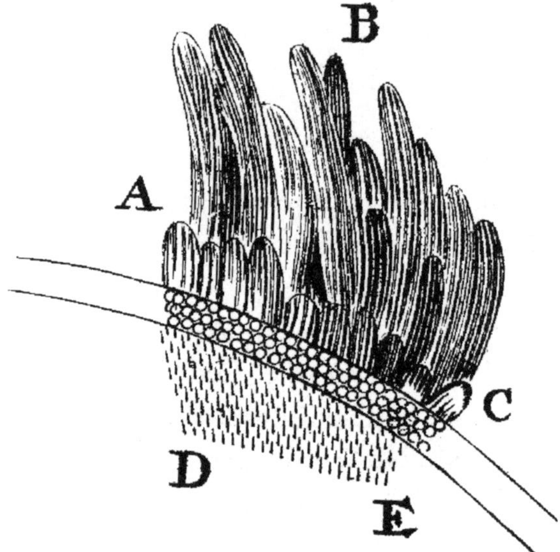
B
A
C
D
E

Introduction: Lost Wonders

> For sale: a microscope with attached mosquito wing. Another with whale meat. Others with a nasal hair, a perch scale, a small tuft of bear's fur. And a lot more such curiosities.[1]

It is 29 May 1747 in Delft and there is a collection of small wonders on offer. They are the work of Antoni van Leeuwenhoek, a resident of the city, who in the course of his long life produced hundreds of glass lenses. Those that were successful he mounted on a small metal plate with a peep-hole. On top of that he fitted another plate, also with a hole, and riveted them together. On the outside, he fitted a needle to hold a small piece of a human, animal or some other natural object, and screws to position the sample exactly in front of the lens. And that was that: he had made a microscope. A very simple one, undoubtedly, but it worked perfectly.

Besides that basic model, with a single lens and a needle, Van Leeuwenhoek also produced many other types of microscope; with a glass tube, for example, instead of a needle, in which he could place blood, milk or other liquids; or with multiple lenses next to each other. Most were small in size, only a few centimetres high and wide, but he also made larger ones. Van Leeuwenhoek used his microscopes to examine everything he could get his hands on: from a bee's sting to gold nuggets to drops of mucus from his own throat. They looked a lot different through a lens than anyone had expected and, for more than a half a century, Van Leeuwenhoek never tired of looking at them.

He shared his observations with the scientific community, which was very impressed. In 1717, a group of professors from Leuven, in modern-day Belgium, wrote an ode to him, praising him as 'the greatest interpreter' of the natural world. In the words of the poem, Van Leeuwenhoek was 'bred from heroes' blood' and with his lenses, 'fragile and small', he had trodden new paths and made new words visible.[2]

But now Van Leeuwenhoek was dead and his microscopes were up for sale. No fewer than 531 were listed in a specially printed catalogue, distributed from London to Berlin and from Groningen to Paris, many still holding the dragonflies' eyes, globules of blood

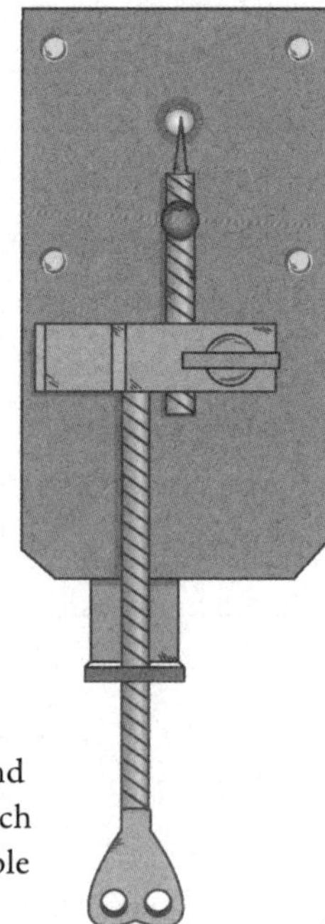

1 A simple Leeuwenhoek microscope with a lens between two metal plates and an adjustable needle on which to place a sample, for example a mosquito wing.

2 Detail of cherubs playing with lenses from the title page of a sales catalogue for Van Leeuwenhoek's microscopes, 1747, etching by Caspar Jacobsz Philips. The one on the right is looking through a special microscope, with three lenses next to each other.

or fragments of diamond that Van Leeuwenhoek had examined through them.[3] A motley variety of potential buyers turned up at the sale, including a salt and soap merchant, a physician, a carillon and organ player, a vicar and an unknown woman referred to only as 'the widow Toornburg'.[4]

The most prolific buyer of the day was Willem Salomons Vlaardingerwout, a lawyer from Delft. He spent more than 77 guilders – almost two months' salary for the average family in his city.[5] For that, he went home with no less than 17 silver and 26 copper microscopes and 12 sets with loose parts, together with a cross-section of the kinds of objects that Van Leeuwenhoek used to enjoy examining: a piece of rock crystal, a fly's brains, the eyes of a large beetle and slivers of ebony.[6]

The lawyer was not well known for his interest in scientific research, which raises the question why he wanted all those microscopes. Perhaps he purchased them for clients from out of town town, who found the journey to Delft long but wanted a Van Leeuwenhoek microscope.[7]

Baker Dirck Jansz Haaxman, a distant cousin of Van Leeuwenhoek's, also bought a number of microscopes. One served as a toy some 75 years later for one of his descendants: as a child Pieter Jacob Haaxman (1810–1888) used it to look at insects and their wings. But he had no idea that his toy was so valuable.[8] He failed to take care of it and it was lost, like many of the other objects sold in 1747.[9] The great majority in fact were lost, which is why only ten are known to have survived, plus one, the authenticity of which is almost, but not 100 per cent, certain.[10]

Fortunately, since the childhood of the younger Haaxman, a lot of information has been gathered about Van Leeuwenhoek, by Haaxman himself, who wrote a book on his discoveries in 1875, and by generations of archivists and academics since then.[11] Their sources include hundreds of letters written by Van Leeuwenhoek, which have much to say about his life and especially his work. In 1923, two hundred years after Van Leeuwenhoek's death, a group of researchers decided to publish a collection of his letters, with explanatory notes. That marked the start of an immensely valuable and rather extensive project, which was recently completed, just in time for the third centenary of Van Leeuwenhoek's death.

In the meantime, historians have unearthed more and more details about Van Leeuwenhoek from archives, and the search goes on. Much new and valuable information can be found on Van Leeuwenhoek specialist Douglas Anderson's website, www.lensonleeuwenhoek.net. Thanks to all this new knowledge and a century and a half of developing insights, we now have a completely different picture of the man than in Haaxman's time.

And now, too, historians are working on new angles. A good example is *Visualizing the Unknown*, a project in which I am involved. Researchers from institutions like the Huygens Institute and Rijksmuseum Boerhaave are trying to understand Van Leeuwenhoek and his contemporaries better by replicating their work. That means not only looking at texts, as historians are accustomed to doing, but making use of original apparatus. In that way, they can find out how men like Van Leeuwenhoek worked, what they saw and especially how they explained to others all the strange and wonderful things they encountered.

Thanks to all the work and ideas of my predecessors and colleagues, I have come to know Antoni van Leeuwenhoek a little better. In the first instance, as a man who could endlessly look with the 'greatest pleasure' at the world his lenses revealed and then enthusiastically – if a little chaotically – tell the rest of the world about it. As a man who made groundbreaking discoveries, but then interpreted them wrongly with great conviction. As the man who examined 'little seed animals' and amoebae and is therefore known as the father of microbiology, but who would be surprised at that accolade. As one of the best observers in history, who never completely belonged among the researchers whose recognition he so eagerly sought.

This book is about a man full of contradictions and the world in which he lived.

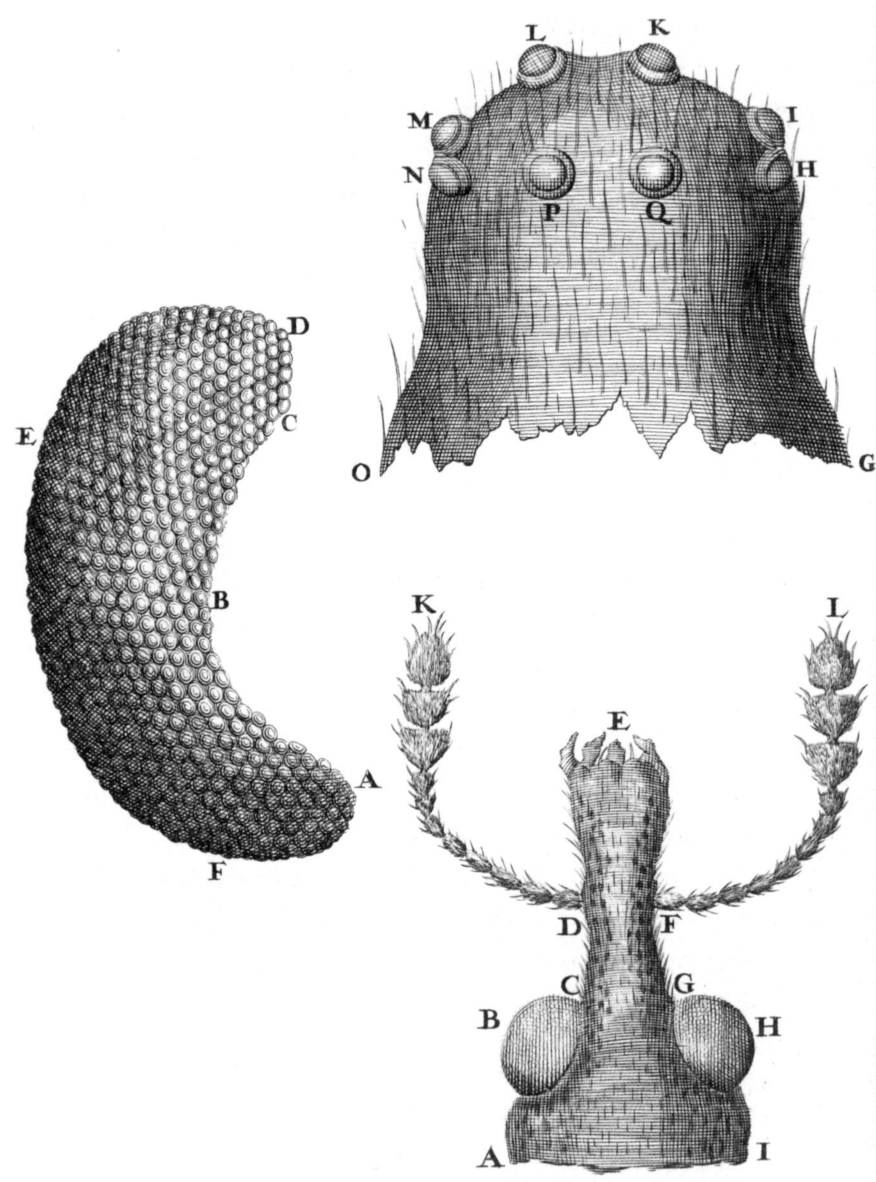
L
K
M
I
N
H
P
Q
D
C
E
O
G
B
K
L
E
A
F
D
F
C
G
B
H
A
I

ONE

Child of Honest Parents

Zacharias Conrad von Uffenbach was in luck. The rich young German was travelling through Northwest Europe in search of knowledge. He visited libraries, spoke to scholars and admired plants, animals and works of art from around the world in the homes of collectors. Now, in December 1710, he was in Delft and was going to visit one of the best researchers of his time, Antoni van Leeuwenhoek.[1] His brother was travelling with him.

They found their way to a narrow house on one of the city's canals. A woman in her fifties, their host's daughter, showed them in. The interior was furnished in typical Dutch style, the walls hung with landscapes, portraits, maritime scenes and images of animals.[2] She led the brothers up the stairs and through a bedroom to a workroom at the front of the house.[3]

There they met the man himself, who was by then almost eighty years old. He looked fit and healthy for his age and his eyesight was still remarkably good, Von Uffenbach wrote in his account of their travels, though he did suffer some pain in his feet.[4] And he was very friendly, allowing the brothers to examine all kinds of tiny wonders through his miniature microscopes: a grain of sand that looked, when magnified, like a beautiful crystal;[5] the scale of a fish, which proved to consist of many layers; a fly's eye, comprised of countless hexagons;[6] and small strands of white material that you could press from the skin of your nose. Some people thought they were worms or maggots but, Van Leeuwenhoek told the brothers, those people were wrong. The strands were the roots of hair that grew on the skin. Von Uffenbach found that an acceptable explanation as, through the microscope, the strands indeed looked like hairs.[7]

Particularly fascinating was his demonstration with a live fish. Van Leeuwenhoek had wrapped the head and body of the fish in a piece of cloth, so that only the tail was visible. He then mounted the whole thing in front of a glass plate, with a microscope on the

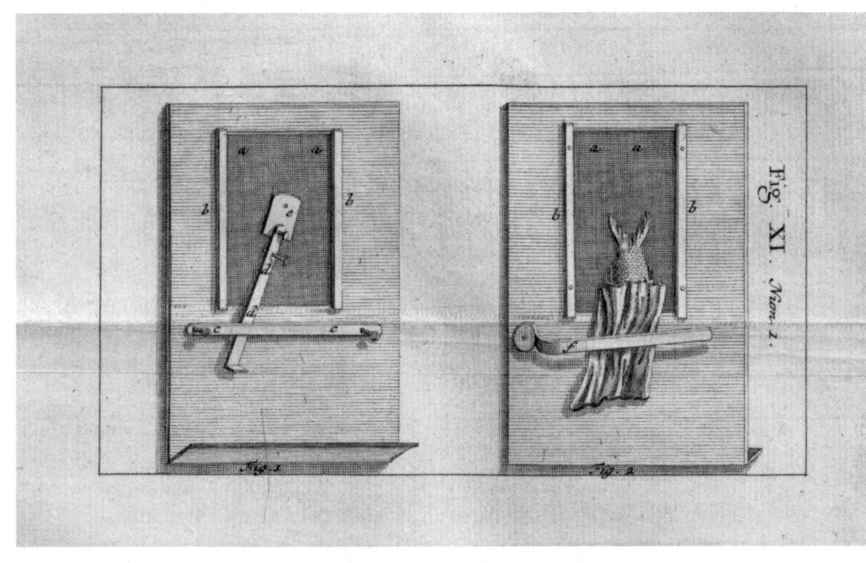

3 Van Leeuwenhoek had found ways of studying the blood circulation of a fish. At first, he had placed the fish in a glass tube and examined them through a lens. But he later found a better solution. This illustration shows the front and rear. It consists of a metal plate with a hole containing a thin plate of glass. On one side (shown on the left) he mounted a movable microscope and on the other a live fish. In that way, he could examine the small details in the tail, including the blood flowing through it.

other side which he could focus exactly on the tail. That enabled the brothers to see the thinnest parts of the tail, and even the blood flowing through them.[8]

But Van Leeuwenhoek had also said strange things, for example about the grain of sand that looked like a crystal. He said that all grains of sand were as beautiful, but Von Uffenbach found that rather a bold claim.[9] And he was dubious about the hexagonal sections in the fly's eye. Van Leeuwenhoek had told him that they were all separate eyes, and that a fly had hundreds of them, perhaps even a thousand. Von Uffenbach thought that was nonsense. That would mean that a fly would have more eyes than the mythical Greek god Argus, whose body was covered with them.[10]

Moreover, Von Uffenbach found Van Leeuwenhoek's instruments rather rudimentary. His apparatus for holding the fish tail, for example, was difficult to use. Von Uffenbach had to place it on his forehead and then look up to see the blood flowing. That went well for a short time, but the position was too irritating to maintain for any length of time.[11] Even worse, they all looked so shoddy. Any researcher worth his salt used nicely finished and preferably decorated instruments. But, Von Uffenbach noted, the metal parts of Van Leeuwenhoek's microscopes were badly made: they were not uniform and the silver was not even pure.[12]

All in all, Von Uffenbach considered Van Leeuwenhoek a good man, very inventive and diligent. But his ideas were badly substantiated and he had clearly not enjoyed a decent education.[13] So why was he so famous?

Fickle finger of fate

It had little to do with his father and mother, who led inconspicuous lives. They were, however, according to a fellow resident of Delft, 'very refined and honest parents'.[14] That refinement was largely due to his mother, Grietge Jacobs van den Berch, who had been born into an influential family in the late sixteenth century.[15] Her paternal grandfather had been an alderman and her great-grandfather mayor of Delft.[16] Her relations had become wealthy as merchants, trading in wares such as beer, a lucrative market in which Grietge's father was also active.[17]

But that prosperity did not last. In the early seventeenth century, times were hard for the brewers of Delft. The price of grain was rising and there was growing competition from neighbouring cities such as Rotterdam.[18] That must have been a source of worry for Grietge's parents, who had together brought seven children into the world.[19] In those difficult years, Grietge's father borrowed the stiff sum of 3,000 guilders from an aunt, an amount that most people in Delft would take twenty years to earn.[20] And then he died, in 1615, probably aged around fifty.[21] The family remained behind with no father and the debt. The calamities continued: the mother also fell ill and died at the beginning of 1616.[22] That left Grietge an orphan at the age of twenty.

Fortunately, her uncles and aunts provided a safety net. And Grietge met Philips, a basket maker, whom she married in 1622.[23] Financially, Philips was a little lower on the ladder than the rich brewers Grietge had grown up among, and his family held no prominent positions on the municipal council. By seventeenth-century standards, Grietge had married below her station, but Philips was by no means a bad match. His father Thonis was also a basket maker.

He had earned enough to buy a few houses, which he rented out or allowed his adult children to live and work in.[24]

Moreover, the merchants of Delft traded in many kinds of goods, which, as many of those goods were transported in baskets, was good for basket makers. And demand for baskets would only increase in the following years, thanks to local producers of earthenware. Around 1600, they had become acquainted with Chinese porcelain and were impressed by the refined bowls, jugs and dishes. Attempts to copy them failed due to lack of know-how, but they did result in a new type of tableware that was a great deal more elegant than the items that they had traditionally produced. That laid the basis for what we now call 'Delft Blue', which, in the course

4 The baskets made by Thonis' father and grandfather were used to carry pottery.

of the seventeenth century, was largely carried to the rest of the country and abroad in baskets made by weavers like Philips.[25]

House in the Oosteinde

Grietge and Philips had seven children, the eldest of whom were girls: Margrieta (1623), Geertruyt (1626), Neeltje (1628) and Maria (1630).[26] On 24 October 1632, they had their first son, the subject of this book, followed by another boy, Jacob (1635), and a girl, Catharina (1637).[27]

Our protagonist was baptized as 'Thonis' in the Reformed Nieuwe Kerk on the market square in the centre of the city.[28] From there it was about 600 metres' walk to the family home, a canal-side house that grandfather Thonis had bought in the Oosteinde.

In their early years, the children probably played in and around that house, and it is very likely that little Thonis regularly saw his father and grandfather at work. Perhaps, once he was old enough, he

5 Thonis and his family lived close to the Oostpoort in Delft.

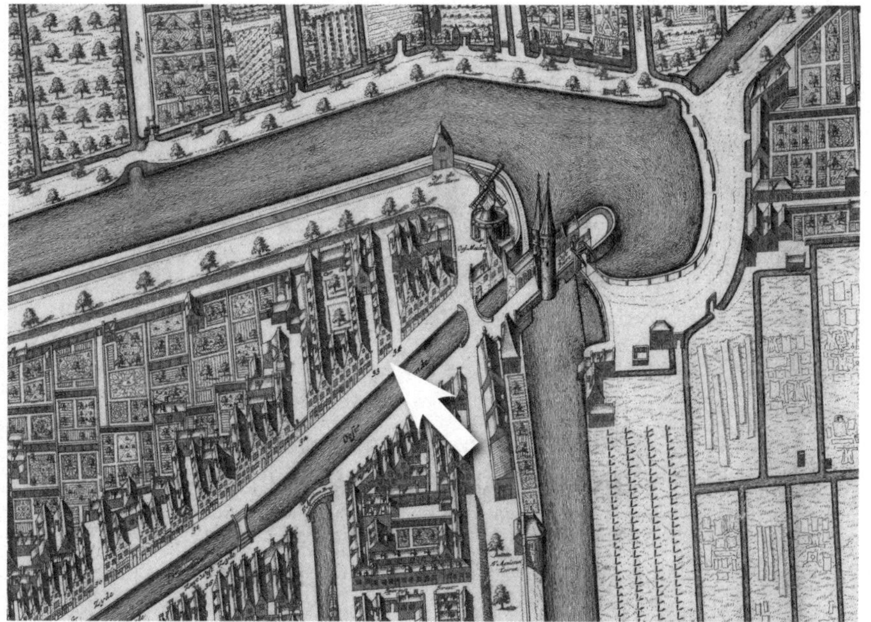

6 The Van Leeuwenhoek family lived in the Oosteinde, shown with an arrow on this street map of Delft.

even helped them with the basket weaving. Though he did not write about that later, it was not unusual for children to do small jobs in the family business. Such early training in handiwork might help a little to explain why Thonis could later do such refined work with glass, metal and tiny animals – though Von Uffenbach would disagree.

The house had no garden, but the children had plenty of space outside to play soldiers or with hoops and tops, with each other or other children in the neighbourhood. Perhaps Thonis used to play with Barbara de Meij, who lived about 200 metres (650 ft) further up the street on the same canal. Three years older than Thonis, Barbara was more probably friends with Thonis's older sisters.[29] But she would play an important part in his life in later years.

If Thonis stepped out of his front door and looked to the left, he would see the Leeuwe-poort, the Lion's Gate. Despite its imposing-sounding name, it was a relatively modest entrance to an alley. The family thus lived on the *Leeuwen-hoek*, Lion's Corner, the origin of Thonis's surname, which was used as of the 1620s.[30]

Crossing the street in front of his house, Thonis would be on the waterside of the Oosteinde. The children liked to fish there. The canals were connected to the River Maas, so that fresh water and fish flowed into the city.[31] If he followed the water to the southeast for about 50 metres (55 yds), he would come to a city gate and a bridge across the canal. That led to small fields where textiles were laid out to bleach in the sunlight, and to gardens where people of the city grew fruit and vegetables. Grandfather Thonis also had a garden there, just outside the gate.[32] In the spring, children hunted for dead-nettles in bloom, and especially for the shiny beetles that were known as goldcrests because of their glossy backs, which young Thonis would research as an adult.[33]

If Thonis stayed in the city and walked from his house 900 metres (about half a mile) to the southwest, he would come to a warehouse that the Dutch East India Company (VOC) had opened on the Oude Delft a year before he was born.[34] The VOC had been transporting Asian spices and other goods to towns in Holland and Zeeland since 1602. These exotic products also reached Delft through the port of Delfshaven, and Thonis was able to learn about the pepper, nutmeg, silk and Chinese porcelain that the ships had brought with them.[35] As an adult, he would use all these products for his research.

A little further along on the Oude Delft was the workshop of spectacle maker Evert Harmansz Steenwijck.[36] Steenwijck ground lenses and even built a telescope, an instrument that was still in its infancy when Thonis was a boy.[37] Perhaps Thonis took a look through one and got an early idea of the potential such lenses offered. But if he did, he did not write about it later.

Stamping letters

Thonis therefore had ample opportunities in his own city to learn about the world and the diversity of life forms that populated it. He also went to school, but by our standards the curriculum was limited. The main aim was to teach children to be good Christian citizens.[38]

Some of the city's children even went to an 'infants' school' from as early as two years old, where they learned how to pray from a schoolmistress – or sometimes a schoolmaster. They also started to learn the catechism and the alphabet so that they could later, as good Protestants, read the Bible and other edifying texts.[39]

We do not know whether Thonis also started school so early. It is quite possible that he skipped infant school and went straight to primary school, known as the Nederduitse school because they learned Dutch rather than Latin. Children usually started at these schools when they were five, but some were younger or older.[40] Here, too, the children started with the alphabet. That was not taught with any great enthusiasm. The children started with A and ploughed on until they could read all the letters. Then came numbers.

Learning to write was not a priority.[41] Little Thonis first had to make the step from reading individual letters to words and even sentences. Sometimes, the children would take an intermediate step and learn syllables, like ba, be, bi, bo, bu, but the emphasis was soon on serious texts, such as Our Father, or extracts from the catechism and Bible stories like that of King David.[42] Not exactly kiddy language, though in the course of the seventeenth century schoolbooks were published containing rhymes of this nature:

> Beware young folk, for evil
> Even if everyone else does not
> Never let a day go by
> Without doing something good.[43]

Such sentiments were completely in line with the objectives of education, in the eyes of Reformed leaders.

Pupils were given an edifying message of a different order in schools giving lessons on national history. The main schoolbook on the subject was *Spiegel der jeugd* (Mirror for the Young), which told of the 'tyranny' and 'barbaric cruelty' of the Spaniards who had wielded the sceptre over Holland and the rest of the lowlands only a few decades earlier.[44] It described the revolt against the Catholic king Philip II, which led to the formation of the Republic of the United Provinces. Thonis's city of Delft played a role in the revolt, as the leader of the insurgents William of Orange had lived there for a number of years and was assassinated there by a fanatical supporter of Philip. For Calvinist contemporaries of Thonis, the murder was typical of the Catholic sadism that played an important role in their self-image: they had wrested themselves free from all cruelties and now formed a truly pious society.

Once Thonis could read well and had the necessary motor skills, he began to learn to write. He was probably about eight or nine years old and it was a difficult step for him. As pencils and pens were still scarce, the standard writing instrument was a goose feather and he had to practise a lot before he could write easily with one.[45]

The writing lessons were also very expensive for his parents. They had to pay extra for the paper he filled with his letters, for the ink and for the schoolmaster who kept the points of the quills sharp. That's why some children left school before they reached this stage, or once they could write their names half decently.[46]

But many parents in Delft found education important and, in the course of the seventeenth century, the number of people in the city who could write increased. That was also to be seen in the Van Leeuwenhoek family. Grandfather Thonis, who was born around 1570, was unable to sign his will in 1643 and marked it with a cross.[47] A few decades later, his grandson would exchange hundreds of letters with great intellectuals in neat handwriting.

Pupils like Thonis used letter-writing books to practise how to start a letter politely (for example, 'My Dearest Sir'), make cordial comments ('Noble Sir, I wish to inform You humbly and with the best of intentions as follows') and finish courteously ('with cordial greetings, I remain Your humble servant').[48] The books contained examples that would prepare the children for their future correspondence with their families, business associates and – in Thonis's case – an international network of scholars.

Arithmetic with Bartjens

Knowledge of and skill in arithmetic was not a requirement to be a good Christian, but it was to conduct business or run a company. Therefore some parents paid schools a little extra to teach their children to add, subtract, multiply and divide.[49] Thonis, too, was taught arithmetic and he clearly knew a great deal more than simply the basics since, as a grown man, he was able to make complex calculations.

There is a good chance that, for the advanced arithmetic lessons, his teacher used *De cijfferinghe*, a book by schoolmaster Willem Bartjens, originally published in 1604. Generations of pupils used the book to learn the standard methods for the kinds of calculations necessary for trade and industry.

Bartjens posed arithmetical problems like '2 pounds of butter costs 7 stuivers. How much does 10 pounds of butter cost?'[50] Not a very difficult sum, but Bartjens presented a standard method for solving it. He called it 'the rule of three': you take the last number (10), multiply it by the middle one (7) and then divide it by the first (2), and there you have the solution: 35 stuivers.

Perhaps the reader finds this method a little strange and would have solved the sum in a different way. For example, 10 pounds of butter is 5 times as much as 2 pounds of butter and therefore costs 5 times as much: $5 \times 7 = 35$ stuivers. That is an instinctive approach, and anyone with basic arithmetical skills can see that it is correct.

For Bartjens, however, it was not about insight, but rules that could be used in all kinds of circumstances. For example, doing calculations with the bewildering mix of different currencies or measurements used by merchants from different cities. Then they might find themselves having to work out, for example, how many Antwerp ells of fabric equal 87.5 Frankfurt ells, if 10 Frankfurt ells are the same length as 7 Viennese ells and if 35 Viennese ells are equal to 24 ells from Lyon, and 3 Lyon ells are the same as 5 ells from Frankfurt.[51] That was a compound version of the kind of sum that could be solved with the rule of three. Pupils had to apply the rule several times in succession to arrive at the correct answer of 70.

Today's secondary school children learn to solve such problems using an equation with an unknown (x), but when Thonis was at school that method was still being developed.[52] So arithmetic lessons still worked with numbers, making the sums a little convoluted to our eyes.

Babies and the plague

At that time, biology did not exist as a subject at school, and children learned about living things in their free time: when they were fishing, rambling through the gardens around the city or in their daily lives at home. Mother Grietge gave Thonis a little brother and sister, and other women in the area regularly had babies too. Thonis could see how they were made by watching dogs in the street, or rabbits at the breeders. And, given that the seventeenth century was not an overly prudish time, he probably heard about how that worked with people, too. That making a baby involved having sex, and male seed, would have become clear to him soon enough. But as to the question how life exactly originated, opinions had been divided since antiquity. Perhaps the male seed contained the core of the tiny new human being, and the womb offered only a warm, protective and fertile shelter in which it could grow. Or perhaps there was also female seed that was released during orgasm, or it was the bloody substance excreted during menstruation – opinions varied on that point, too. In any case, the proponents of the 'two-seed theory' assumed that a mixture of the seeds was the basis of the new life, together with the soul, which came from God. That was clear to everyone.

Many of those involved in the debate were not completely satisfied with the technical part of the theory, and during Thonis's life researchers would attempt to formulate a more convincing explanation. Thonis himself would become one of them.

*

Of all the babies born in Holland at that time, an alarming 25 per cent died before they were a year old, and another 20 per cent before they were five. After that, the greatest dangers were past, but a few per cent still died in the years that followed, so that around half reached the age of eighteen.[53]

Thonis experienced this at first hand. His younger brother Jacob died before his eighth birthday and his sister Maria when she was thirteen.[54] Another sister, Neeltje, did reach adulthood but died at 22.[55] Significant causes of all these deaths were 'pestilences', which would come in waves.

Today, we know that the disease we call the plague is caused by the bacteria *Yersinia pestis*, but in the seventeenth century the microbe had not yet been discovered. Contemporaries, however, used the term more expansively than we do. Historiographers have tried with hindsight to categorize historical epidemics in modern terms, but that is often extremely difficult. Nor is it important to understanding Thonis's world and how he saw it. What matters is that, every now and then, the city in which he lived was stricken by a medical disaster. In both 1635 and 1636, for example, when he was about three years old, three hundred more people than usual died in Delft, an upsurge owing to the plague.[56]

It is very probable that Thonis's parents saw such an epidemic as a punishment from God, as many people at the time agreed on one thing: if humanity sinned, the wrathful Lord would send disease and pestilence.[57] They would also see natural causes, as divine intervention did not exclude other factors.

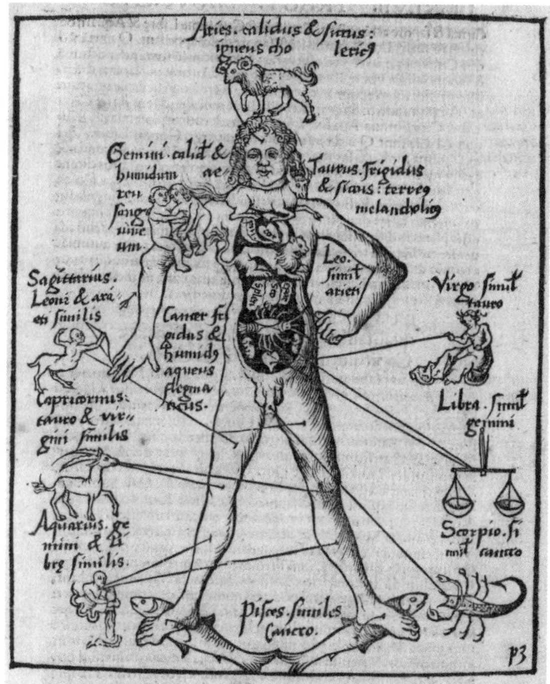

7 A 'zodiac man' illustration showing how body and zodiac signs were connected.

Those other factors came, for example, from the heavens. It was widely accepted that the stars and planets had an influence on people's health. The 'macrocosmos' of celestial bodies was closely connected to the 'microcosmos' of the human body. The signs of the zodiac were particularly important, as each sign affected the health of a part of the body: Cancer had an impact on the chest, Aquarius on the lower legs and so on, as can be seen from illustrations of 'zodiac men', which offered a useful overview of the link between the body and the heavens.[58]

Closely related to this system was the theory of the four elements, or principles: earth, water, air and fire. These elements were present everywhere on Earth, and thus in people too. All four had basic properties: they were warm or cold, dry or wet, each in their own combination, as shown in the table below. The same combination was also apparent in the four fluids or 'humours', which were essential to medicine. The body was seen not so much as a collection of organs, but as a mixture of the fluids: blood, yellow bile, black gall and phlegm.[59]

Elements, bodily fluids and how they are connected

Earth	*cold and dry*	black gall
Water	*cold and wet*	phlegm
Air	*warm and wet*	blood
Fire	*warm and dry*	yellow bile

Ideally, the bodily fluids were in balance, which varied for each individual according to age, sex and so on. Small children, for example, were conspicuously wet, and boys and men were usually warmer than girls and women.[60]

Unfortunately, that balance was continually under threat. Then an individual would be too cold or too warm, too wet or too dry, and consequently not in the best of health, or seriously ill.[61] Because their fluids were out of balance, children – and boys in particular – were especially susceptible to virulent diseases like smallpox, which periodically took many lives, particularly those of the youngest.[62]

Adults were less vulnerable, but they could also become unwell in all kinds of ways. That could be caused by the macrocosmos. For example, by the cold, wet moon, which, as the nearest celestial body, influenced not only the tides but the flow of bodily fluids.[63] Or it could again be caused by the zodiac, where the signs were not only linked to parts of the body but to the elements and their properties, and therefore with bodily fluids: Pisces, the fish, was for example related to water, and Sagittarius, the archer, to fire.

Other causes of sickness lay closer to home. There was the weather, which could be warm, cold, dry or wet; the changes of the seasons; excessive or insufficient physical activity; and an overly copious or too meagre diet. A balanced diet and a routine of physical activity were basic requirements for balance between the bodily fluids. When health problems did arise, a sensible person would enquire whether it was time for bloodletting or purging. These could restore the balance of the bodily fluids, but only if the circumstances were right. If the moon, and especially the full moon, was in Cancer, a doctor would advise against bloodletting in the case of a patient with chest problems. Given the link between the zodiac sign and the body part, such treatment could have undesired consequences.[64]

Yet even those who followed such advice could still become seriously ill, for example, due to malign fumes from water or the soil, or to seeds in the air that carried disease and spread pestilence.[65] They could be combated by airing rooms, burning herbs or – in serious cases – firing off cannons and cleansing the whole living environment with vinegar.[66] If that all failed, healing herbs and other medicines would become necessary.

Away from Delft

Perhaps Thonis became acquainted with these ideas when members of his family were sick. That probably happened quite often, and was sometimes very serious. Besides his sister and little brother, he also lost his father at an early age. Philips died in 1638, when Thonis was still only five years old.[67]

The family could stay on the Oosteinde, as the house was owned by grandfather Van Leeuwenhoek. But mother Grietge would nevertheless have found it difficult after the death of her husband, and certainly in a practical sense, as seventeenth-century society was not set up for single parents. Widows and widowers soon sought a new partner, and Grietge remarried within three years. Her new partner was Jacob Jans de Molijn, a tax collector and painter who had already lost two wives.[68]

Jacob had four children from his earlier marriages, three of whom were still so young that they probably lived with him.[69] That made it quite crowded in the composite household, which may be why, as Delft historian Reinier Boitet noted shortly after Thonis's death, the latter was sent to boarding school in Warmond, a little over 25 kilometres (15½ mi.) to the north of Delft, close to Leiden.[70]

He may have attended the Latin school run by Cornelis Loveringh, which offered board to young schoolboys.[71] That would have been remarkable, as the Van Leeuwenhoek family had no connections in Warmond. Moreover, Loveringh was a Catholic, raising the question of why a Protestant family living in an area where there were ample Protestant schools available would choose a Catholic tutor.[72]

There were Latin schools for children – boys – from nine years old, but most of the pupils were a little older. They would build on what they had learned earlier. History lessons were more inspired by the Bible and they could also practise their calligraphy.[73] But the teaching focused on the classics, preparing its pupils for university. They learned Latin and Greek, a prerequisite for university students. And to give them a feeling for style, they read the works of great writers, like Cicero. As there was much debating at universities, the pupils also received lessons in rhetoric and logic, and lastly in drama and music – good for their expressive skills and cultural refinement.[74]

If Thonis did indeed attend one of these schools, it was by no means a resounding success. In his later letters, he would note that he could speak no foreign languages. So any Latin or Greek he may have learned was long forgotten. The same applies to instruction in style or rhetoric; his later letters showed little rhetorical flair. And he was certainly not the most lucid author of his time.

*

It remains uncertain whether Thonis went to the Latin school in Warmond. But, according to Boitet, after Warmond he went to Benthuizen, which is a little more logical. There, some 16 kilometres (10 mi.) to the northeast of Delft, lived his Uncle Cornelis, his mother's brother, who had left the declining beer brewers' branch and become a sheriff. With his uncle, Thonis acquired legal and administrative experience, perhaps with the intention of one day following in his footsteps.[75]

But that was not to be. According to Boitet, by the age of sixteen, Thonis was living in Amsterdam, 'with a prominent textile merchant'.[76] That merchant was William Davidson, a successful Scottish trader with excellent connections, including the English royal family.[77] We know this information from a document of 1653 in which Davidson gave considerable responsibilities to 'Antonij Leeuwenhoeck' (so no longer going by Thonis), then only nineteen years of age. Antoni did bookkeeping and was permitted to conduct bank business for his boss.[78]

The responsibility suited him. As he was to prove in later years, the boy was accurate and loved working with numbers. It was therefore logical that Antoni would continue to work in the textile trade. But that was to be in his own city of Delft.

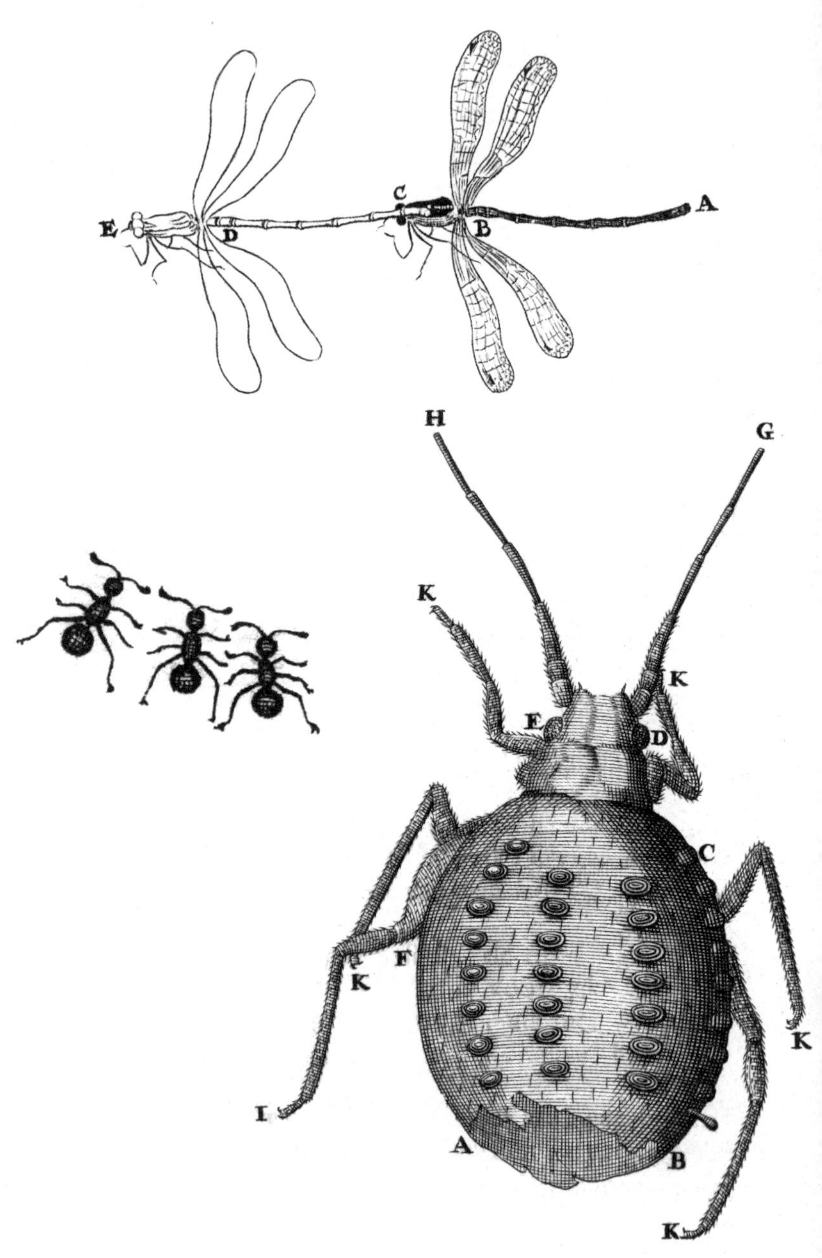

A
B
C
D
E
H
G
K
K
F
D
C
F
K
K
I
A
B
K

TWO

Meeting the Minuscule

A man of the world: that's how Antoni had himself portrayed around the age of fifty. He dressed himself in style for the occasion, with a full wig of curls, an ultra-thin moustache and a shiny housecoat. A globe, a quill and a pair of compasses symbolize his studious side, and on the extreme left of the painting lies written proof of his status: a beautifully inscribed parchment with a hefty seal. Most of the words are quite obscure, but the artist, fellow resident of Delft Jan Verkolje, made a few of them clearly legible. *Caroli Secundi*, for example, Latin for Charles II, the king of England, Scotland and Ireland.

The parchment was not from Charles himself, but from the Royal Society in London, of which he was the patron. The Society was founded in 1660 by a group of gentlemen from well-to-do circles who wished to conduct scientific research. Many of them had attended the highly esteemed universities of Oxford or Cambridge, and the most prominent among the original members was Robert Boyle, son of the wealthy Earl of Cork, who had sufficient funds to spend all his time on research into natural phenomena.

And in 1680, so the parchment tells us, this eminent society had made basket-maker's son Antoni van Leeuwenhoek a fellow. He was rewarded because he was a 'highly intelligent' man who had made 'very ingenious observations': observations that the members of the Royal Society were able to confirm only after considerable efforts.[1]

The portrait shows how proud Antoni was of his membership. But the elegant way in which he had Verkolje immortalize him is only a small part of his story.

The sorrow of a young couple

When the English scholars made him a fellow, Antoni had been living in his native city again for many years. There in 1654, at the age of 21, he had married Barbara de Meij, a girl who had lived a little further up the street when he was a boy, alongside the same canal.[2]

Barbara's late father had been a textile merchant and had left her a considerable inheritance.[3] Antoni used the money to set up a business in the same branch. He sold wool, silk and other fabrics, as well as accessories like ribbons and buttons.[4] He did that from a canal-side house that he bought shortly after his marriage, where he would live for the rest of his life and where Von Uffenbach would later visit him. It was in the Hippolytusbuurt and was called Het Gouden Hoofd.[5]

The house has since been unrecognizably rebuilt and combined with the neighbouring corner house. What little remains, including the foundations, are now part of Hippolytusbuurt 1. Two houses further along there is a plaque claiming that Antoni lived there, but that is incorrect. Readers looking for a historical Van Leeuwenhoek experience in Delft really have to be at number 1, even though there is little of the seventeenth century left to see.[6] They can go in: at the time of writing it houses a restaurant serving breakfast and lunch.

*

With his house, Antoni was well set up in material terms, but in other aspects of his life, during these years, it seems as if one catastrophe followed another. The misery started on 12 October 1654, when an explosion in a secret gunpowder store in the north-east of Delft took around one hundred lives, or perhaps many more. A few hundred houses were so badly damaged that they were beyond repair.[7] We do not know whether Antoni lost friends, neighbours or other acquaintances, as in the hundreds of his letters that have survived he makes no mention of what has become known as the 'Delft thunderclap'.

Nor did Antoni write of sorrowful events in his own household. Like many parents of their time, he and Barbara had to bury several of their children. A year after their marriage, they had a son, whom

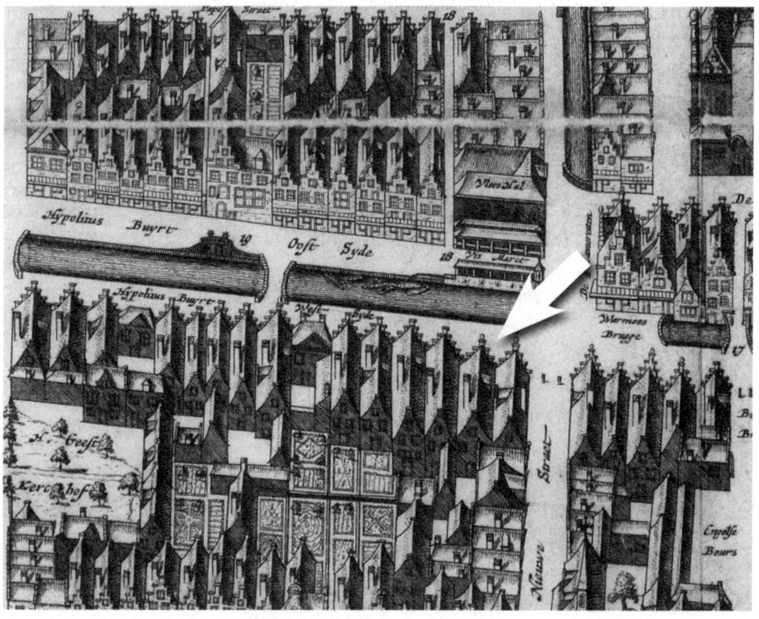

8 From 1655, Antoni lived in the Hippolytusbuurt, in the second house from the corner of the Nieuwstraat.

they named Philips, after his grandfather.[8] The baby died at a little more than a month old.[9] He was followed by a new baby, also named Philips, but he also died after only a few weeks.[10] A third Philips lived to celebrate his second birthday, but died shortly afterwards.[11] And Barbara and Antoni had to bury their daughter Margrieta, named for her grandmother, a week after her baptism.[12]

That Antoni did not speak of his late children in his letters by no means meant that he was insensitive. He and Barbara will certainly have mourned their lost sons and daughter. But in the seventeenth century, people did not write as much about emotional matters as we do nowadays, and certainly not in the kinds of letters from Antoni that have been preserved, which mainly served to exchange knowledge and not to discuss emotions.

One child did appear in the letters, because she survived. Maria was born in 1656 and proved strong enough to withstand all childhood diseases. She would remain with her father for the rest of her life, in Het Gouden Hoofd.

The numbers man

Between all the births and deaths, Antoni's work took a different course – one that was more closely aligned to the talents he had displayed in Amsterdam while working for William Davidson, and which would become very useful in his later research with microscopes. In 1660, he was appointed *camerbewaarder*, a kind of all-purpose assistant to the sheriffs, who were responsible for governance and law. They did that from the town hall, which was a stone's throw from Antoni's house. Whenever the sheriffs came together, it was his task to ensure that everything went off smoothly. He would open the meeting room and close it again at the end of the day, clean it, light the fire and make sure there was enough coal to keep it burning. Or he would hire someone to do some or all of these jobs. He probably also had organizational and administrative tasks, such as writing letters.[13]

At first, he earned 260 guilders a year, plus 54 guilders for any additional expenses. In 1664, his salary was raised to 400 guilders, with the same allowance for expenses.[14] These were substantial sums at a time when a tradesman would earn some 200 guilders a year.[15]

Despite his adequate earnings, Antoni continued his textile business for a few years, but the balance gradually tipped in favour of his work at the town hall. Besides his tasks as a *camerbewaarder*, the city councillors gave him work to do that was better suited to the legal, administrative and financial training he had received in Benthuizen and Amsterdam. As the years passed, he acquired a significant position in the financial affairs of the city.

It most probably started with simple tasks like issuing summonses or questioning witnesses. *Camerbewaarders* could earn a little extra by performing these tasks.[16] Gradually, he took on more and better-paid jobs. In 1666, Antoni became one of two general district supervisors in Delft.[17] He and his colleague were the point of contact for sixteen supervisors, each of whom was responsible for keeping order in their district of the city. If problems arose, Antoni would discuss them with the city mayors, of whom – unlike now – there were a number in office at the same time.[18]

The city councillors were apparently satisfied with his work as, in 1667, they gave him another task to fulfil – an investigation that would prove very time-consuming. The subject of the investigation was Sijmon Bourbon, the black sheep of a wealthy Delft family. Sijmon was the same age as Antoni and had inherited thousands of guilders in shares and other property. That generated so much money that, under normal circumstances, he could have lived very comfortably. But, in his family's eyes, Sijmon did not behave normally.[19] He spent his inheritance so quickly that he soon ran into financial difficulties. He had spent a large part of his money in taverns, and owed one innkeeper 100 guilders, half a year's salary for many of his contemporaries.[20]

His family were concerned about Sijmon, or their good name, or perhaps both. Financial unreliability was bad for the family's status and thus for the trade from which part of the family made its living.[21] So his relations appealed to the highest court in Delft, requesting that they place the wasteful Sijmon under trusteeship.[22] That task was entrusted to Antoni, which required him to put Sijmon's financial affairs in order.

It took him a year to complete the investigation, listing all expenditure and debts to the nearest cent and divided into clear categories. He even recorded all the expenses he incurred in conducting his inquiries, such as 33 sheets of paper for all his auditing notes.[23] Moreover, he ensured that all creditors received their money, so that the Bourbon family could once again breathe freely. He had averted a financial catastrophe.

Sijmon would no longer disgrace his family as, in the late spring of 1667, following in the footsteps of many other black sheep from wealthy families, he boarded ship for the far-off East Indies. Antoni also took care of the arrangements, making sure that Sijmon left Delft with freshly washed and ironed linen, and accompanying him to Amsterdam by barge. A friend then travelled further with Sijmon to the island of Texel, from where he sailed for Java.[24] He would never return. On 26 July 1668, Sijmon died near Makassar, aged 33.[25]

*

After his success with the Sijmon Bourbon affair, Antoni would deal with many other similar cases, including that of a man who, after his death, was to become world-renowned. Artist Johannes Vermeer, a fellow resident of Delft, was a few days older than Antoni. Both had been baptized in the Nieuwe Kerk, and their names appear on the same page of the baptism book.[26] But, while Antoni would live to be ninety, Vermeer died in 1675 at the young age of 43. He left many debts, and it fell to Antoni to ensure that his creditors received their money.[27]

Some have concluded from the fact that Antoni dealt with Vermeer's affairs that the two must have known each other.[28] It is a tempting idea: two talented figures from Delft spending time together and perhaps even discussing innovative ideas. But Antoni took on these tasks because he was known at the town hall and because of his experience of working with numbers. The fact that he dealt with Vermeer's estate therefore says very little.

Yet it is quite possible that the two men encountered each other in Delft, which was then still a moderately sized town, even more so because they had mutual acquaintances. Antoni's stepbrother Jan Jacobs de Molijn, for example, was a painter, as was his father, Jacob. He was also married to Antoni's eldest sister, Margrieta, and the stepbrothers/brothers-in-law stayed in touch after both had left the family house.[29] Like Vermeer, Jan was a member of the guild of St Lucas, which painters were required to join.[30]

Because of the reasonable likelihood that the two men knew each other, there has been speculation that Antoni was the model for two of Vermeer's works portraying thinking men: *The Geographer* and *The Astronomer*.[31] But the faces in Vermeer's paintings bear no resemblance to Verkolje's later portrait of Antoni. There is no further evidence to support the idea.

Medical input

While Antoni was working on his new career at the town hall, another change in his life presented itself. That was probably related to the arrival of Regnier de Graaf, a physician who moved to Delft in 1666. De Graaf had attended a number of universities – at Leuven, Utrecht, Leiden and Angers – and had also been to Paris, one of the most prominent intellectual centres of the time.[32]

During his studies, De Graaf had learned much about anatomy and, at Leiden, had taken part in innovative research. That was by no means common at the time, as universities focused mainly on education. Research was usually conducted elsewhere, for example by associations of scientists like the Royal Society in London. But the medical faculty at Leiden was an exception. Many innovative scholars worked at the faculty and one in particular was of great interest to De Graaf. Franciscus de le Boë Sylvius (1614–1672) was fascinated by iatrochemistry, a new branch of medicine derived from alchemy. Iatrochemistry was based on the notion that the human body functioned on the basis of chemical reactions, particularly between acids and their chemical opposites, alkalis. If they came into contact, they caused crucial 'effervescences'.[33]

In 1664, when De Graaf was studying in Leiden, Sylvius was researching the pancreas, an organ close to the stomach and the gall bladder that releases fluids into the small intestine. Sylvius had concluded that the fluids were acidic and reacted in the intestine with alkaline fluids from the gall bladder, producing effervescences that helped digest food.[34] Sylvius charged De Graaf with proving that the pancreatic fluid was acidic, and the latter began experimenting with dogs. He cut open the abdomen of living dogs and connected the pancreas to a bottle using the shaft of a duck feather, so as to capture the pancreatic fluid. After a number of failed attempts, one dog stayed alive long enough for De Graaf to tap off sufficient fluid.[35] To stop the dog from whining pitifully, he cut a slit in its windpipe so that it lost its voice.[36]

Once the fluid was in the bottle, De Graaf and Sylvius could taste it. The taste varied quite significantly, but both decided that it was predominantly acidic. That is a little strange, as the fluid is in reality alkaline. What probably happened was that Sylvius, and through him De Graaf, was convinced in advance that the fluid would taste acidic, and the experiment 'proved' the accuracy of this assumption.[37]

De Graaf's experiment shows us clearly what kind of medical research was conducted at the time, but his attention to the pancreas is also important for another reason. There was a theory that the fluid did not come from the pancreas itself but from the nervous system, and De Graaf wanted to disprove that notion. So he examined nerves in search of an aperture through which the fluid might be able to flow. But his records show that he could not see one. No matter how closely he looked, he was unable to find an opening, even using the most accurate microscopes.[38] That De Graaf was clearly familiar with microscopes in the 1660s suggests that it is quite possible that he showed them to Antoni during his years in Delft.

Ancient lenses

When that happened – if it happened – the microscope had already been in use for some time. And the most important component, the lens, was already ancient. In the distant past, people had observed that objects seen through a transparent, curved medium – such as a drop of water – looked bigger. With a little imagination, you could say that those water droplets were the first lenses.

In antiquity, various civilizations were able to make lenses, and we have clear examples from the Romans.[39] One was found in the house of an engraver in Pompeii, which had been inundated with lava after Mount Vesuvius erupted in AD 79. That raises the question of whether Roman craftsmen used lenses, perhaps to make cameos, finely engraved gemstones with relief portraits of emperors or other images. Magnifying lenses would have been very useful for this kind of delicate work, but unfortunately the surviving sources do not tell us whether they were used for this purpose.

We do know for certain that Romans used lenses to help them read. They would lay a lens – for example, a glass ball filled with water – directly on top of the text to make the letters bigger.[40] These 'reading stones' would be used until well into medieval times, but they had their drawbacks. They often contained specks of dirt or air bubbles, which limited their effectiveness.

A turning point came in the late thirteenth century, when Venetians developed a new type of exceptionally clear glass. That enabled them to make relatively small lenses, which could be mounted in pairs in a frame that you could put on your nose.[41] That marked the birth of spectacles, though as yet without side arms, which would not be developed for another few hundred years.

Spectacles gave each eye an extra lens, and (more or less) with the appropriate strength. They also had one great advantage over a reading stone: because spectacles did not lie directly on top of the text, they could also be used for writing. That was good news for people who made a living by writing – a growing group in the late Middle Ages.

The next step came about two hundred years later, at the end of the fifteenth century. Until then, lenses were often ground using

9 In the late 16th century, Joris Hoefnagel drew insects so accurately that it seems conceivable that he used lenses to see them in detail.

a flat grinding disc, but now lens-makers came up with the idea of using hollow bowls lined with abrasive material. In that way, they could produce lenses with the right curvature.[42]

That was a great step forward for anyone who was short-sighted, and for anyone who needed lenses for other reasons – such as textile merchants, who used lenses to examine the thread in fabrics close up to check the quality, and perhaps also artists with an eye for small details. Joris Hoefnagel (1542–1600), for example, drew many plants, insects and other small animals in the greatest of detail, leading to conjecture that he used lenses to study them from very close up.[43]

Lenses ground in bowls had one significant disadvantage, however: they were globe-shaped, or at least partially globe-shaped, and that meant that they always produced a distorted image. Ideally, a magnifying lens focuses all the incoming rays of light into one point, but with these kinds of lenses the rays are always slightly dispersed,

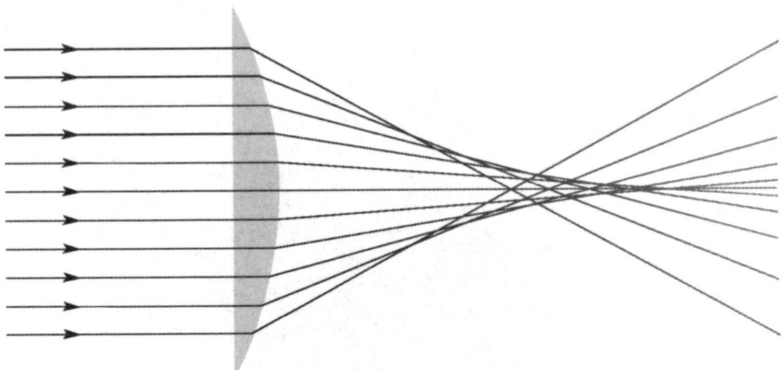

10 The curved side of this lens is part of a globe. Lenses like this never concentrate the incoming rays of light into a single point, as can be seen on the right.

so that the image is not completely sharp. The further from the centre of the lens the rays pass through it, the worse this problem is.

In addition, white light is made up of all colours of the rainbow and a convex lens bends them all slightly differently. Anyone looking through a convex lens will therefore see an image where the

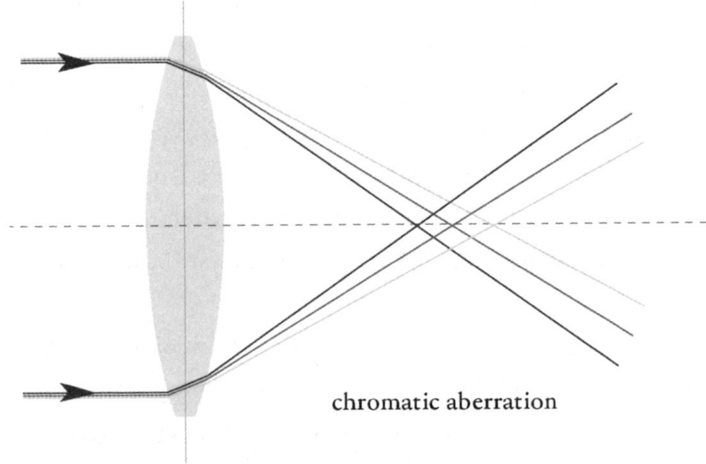

11 The various colours in white light all curve at slightly different angles when they pass through a lens whose surfaces are part of a globe.

colour is distorted. That effect is also more marked with rays that pass through the lens further from the centre, because that is where the colours are dispersed most widely.

Without being aware of this theory – the properties of light were not understood until later – lens-makers tried to solve the problem of blurred images. When those efforts proved futile, one resourceful mind decided around 1600 to mask off the exterior edges of the lens, thus reducing the cause of the blur.[44]

It was a smart move, and was particularly popular among stargazers. With a partially masked-off lens, they could see much further into space. If they affixed the lens to one end of a tube, with a concave lens at the other end; that made the image even better. They also discovered that if they used two convex lenses in a tube, it magnified the image even more. These tubes with lenses are what we now call 'telescopes'. Astronomers used them to gather evidence that Earth orbits the Sun, rather than vice versa, which had long been the conventional wisdom. Around 1600, the theory of the orbiting Earth was still controversial, but it was soon to become mainstream among astronomers.

The invention of the telescope also signified the birth of another scientific instrument. Anyone who pointed their tube with lenses not at the stars but at what was in front of their noses had a microscope in their hands. This alternative purpose called for a slightly different design, and microscope users experimented with different lengths of tube and numbers and types of lenses. That enabled them to build an instrument that allowed them to study the natural world more closely than Hoefnagel had done.

Among those to do this were the members of the Accademia dei Lincei in Rome, which was founded in the early seventeenth century.[45] It was a scholarly community whose members occupied themselves with natural research and artistic pursuits like literature

12 Pope Urban VIII received the *Melissographia* in 1625, depicting three bees as seen through a microscope, as a tribute to him and his family. The engraving was the work of Matthias Greuter.

and music. That was how things were in those days: intellectuals did not focus on one specialism but developed themselves in broad terms, in scientific knowledge and in the arts.

The most famous member of the Accademia was Galileo Galilei, an enthusiastic early user of lens tubes. He mainly used them to look at planets and stars, but amended his instruments so that he could also observe flies. Under his microscope, he told a contemporary, they looked as big as lambs and were covered in hairs. He also said that they had nails at the end of their legs, which enabled them to walk on vertical glass.[46]

Two other members of the Accademia, Federico Cesi and Francesco Stelluti, made serious work of microscopic observations and especially of a subject that gained popularity in 1623. In that year, a member of the noble Italian family Barberini was elected as Pope Urban VIII. The new pope was known for his love of art and knowledge, and now that he was in a position to hand out money and jobs, Cesi and Stelluti decided it was time to develop a close relationship with him.[47] They did that by investigating bees, which appeared on the Barberini family's coat of arms. Stelluti in particular observed the insects – through a microscope. The two scientists commissioned an artistic engraving of Stelluti's observations, entitled *Melissographia*, which they sent to the pope in 1625. It was the first publication that we know of that was made on the basis of microscopic observations.[48]

On a different Earth

In the Netherlands, Constantijn Huygens (1596–1687) was an early fan of the microscope. A diplomat and art-lover, he had bought one in London in 1622.[49] It took him a while to get used to it, he would write later, as 'people who do not know the instrument complain at first that they see nothing, because they see things that they have never seen before. But very soon they cry with delight that their eyes are observing the most unbelievable wonders.'[50] The lenses showed Huygens how magnificently the Creator had put everything together: 'It is really as though you are standing in front of a completely new theatre of nature, on a different Earth.'[51]

Huygens passed on his enthusiasm for art and scientific research to his sons Christiaan and Constantijn Junior, and they too set to work with lenses. Sometimes they made them themselves and sometimes they bought them from specialists, like engineer Johan van der Wyck (1623–1679). In 1654, Christiaan bought a 'magnifying glass' from Van der Wyck and encouraged his brother to take an interest in the latter's microscopes, through which, according to Christiaan, you could see worms in cream, in flour and in the meat of a hare.[52]

At that time, Van der Wyck was living in Delft, the city to which Antoni had just returned after his years in Amsterdam. And Van der Wyck's lenses were renowned far beyond Holland.[53] There is a good chance that Antoni heard about him in those years. He may even have bought a thread counter from him, a device that textile merchants used to check their merchandise. Perhaps he saw microscopes while he was at Van der Wyck's and took a look through one. It is all possible, but it is also all speculation, because he never wrote anything about Van der Wyck, just as he never mentioned lens-maker Evert Harmansz Steenwijck, who also lived on the Oude Delft when Antoni was a child.

Blood and milk

In any case, in his young years, Antoni had ample opportunities to become acquainted with lenses. But that was not enough to become an influential microscopist. That also required substantive ideas, at a level that Antoni had not encountered at home or at school.

In that respect, Regnier de Graaf appears to have played a role. In Delft, he worked as a physician and also continued to conduct experiments of the kind he had done with Sylvius in Leiden. And, in the late 1660s, Antoni was allowed to watch at least one of them.[54]

The experiment concerned blood and milk and, as with De Graaf's research in Leiden, dogs were once again the victims. First, Antoni recorded, De Graaf cut open one of the dog's veins and tapped it off until the animal became weak.[55] Then he opened a vein of a second dog, which lay close to the first, and set up a transfusion: through the shaft of a bird's feather, he allowed the blood of dog 2 to flow into the vein of dog 1. And it worked: dog 1 recovered.

De Graaf then injected dog 2 with cow's milk. His intention was to prove a theory popular among many doctors at the time that, even though they looked different, blood and milk were fundamentally the same.[56] Just as dog 1 had been saved by the blood from dog 2, dog 2 was to be saved by the cow's milk. But unfortunately for dog 2, the theory was incorrect. 'The milk had hardly been introduced into the vein before the dog died,' Antoni noted.[57] It was a disappointing result, but the theme of blood and milk would remain in Antoni's mind and in later years he would investigate the similarity between them in his own way. Not by means of a transfusion, but with a microscope.

And De Graaf did something else interesting in Antoni's eyes. He investigated the reproductive organs of mammals, and of humans in particular, as he was especially fascinated by the question of how babies originated. In that context, he devoted much interest to the small 'testicles' connected to the womb but whose function was unclear. The testicles contained small vesicles, which De Graaf believed contained the 'eggs'.[58] That theory proved correct: the 'testicles' are what we today call the ovaries and the vesicles are sometimes still known as 'Graafian follicles'. During this research, De Graaf apparently used lenses on occasion, but not necessarily to

13 Regnier de Graaf in his study performing anatomical research, drawing by Jan Verkolje, *c*. 1670–73.

magnify objects. While examining oviducts, for example, he used a lens to concentrate light on the small organ.[59]

Antoni was to adopt a different approach. Like De Graaf, he became fascinated by reproduction, but he constantly used his microscopes in his research. That would lead him to radically different conclusions. But more of that later.

Young widower

While Antoni was exploring new avenues, death continued to affect his family. His mother died in 1664 and, in 1666, he also had to bury his wife, Barbara, shortly after his third son, Philips.[60] So, in his mid-thirties, Antoni was an orphaned widower who had lost four children, leaving him the single father of one daughter, Maria.

In these circumstances, he closed his shop – perhaps because he was too busy at the town hall, but also perhaps because Barbara had been indispensable, helping customers and performing other tasks.[61] He also decided around 1668 to travel to England.[62] It was a journey 'out of curiosity', he was to write later.[63] And that was remarkable as, in the seventeenth century, travelling for pleasure or to satisfy one's curiosity was reserved for a select group. They were usually young men – like Zacharias Conrad von Uffenbach – born with a silver spoon in their mouths who, after completing their studies, travelled around at their parents' expense to deepen their general development. They would first visit one or more universities and then complete their intellectual education with a long and expensive tour, preferably visiting prominent scholars. For less well-educated (former) textile merchants and municipal civil servants like Antoni, such trips would normally be beyond their means.

It was also a striking choice politically, given that the Republic of the Netherlands and England had shortly before fought a war. In those years, England was fervently fighting its way around the world to construct an overseas empire from India to New York and beyond. Its ships battled against its rivals, including that other aggressive maritime superpower, the Dutch Republic. That led to regular clashes between the two countries.

Shortly before Antoni's visit, the recent war had ended in humiliation for the English: in 1667, a Dutch fleet had sailed up the River Medway to Chatham, to the southeast of London. Chatham was a naval base and a shipyard where warships were built. The Dutch fleet partly destroyed the base and the shipyard by fire and towed away the English flagship, the *Royal Charles*, across the North Sea.

Antoni thus visited a hostile power, but the political tensions between the two countries appear to have been irrelevant to him. While the Republic and England fought each other militarily and economically, he established contact with British scholars and would later write hundreds of letters to members of the Royal Society. And he was by no means an exception. At a time when European powers competed with each other in all kinds of ways, scholars from different countries maintained close contact with each other. They exchanged knowledge and ideas by letter, in journals or at meetings, so that new insights spread quickly.[64] These included discoveries that Antoni would make only a few years later.

The whiteness of chalk

Antoni may have gone to England to visit his in-laws. Barbara's father had lived in Norwich, and it is quite possible that, after his wife's death, Antoni went to see her family.[65] But perhaps we should take Antoni's own explanation seriously, and believe that he made the journey purely out of curiosity. There were many fascinating developments in England at the time, in research into nature and anatomy in particular and, more specifically, in the field of microscopy. We can safely assume that Antoni had heard of these developments through acquaintances like De Graaf.

The advances in microscopy were largely thanks to the work of Robert Hooke (1635–1703), a clever young man in his thirties who was responsible for experiments at the Royal Society. In 1665, he had published the *Micrographia*, a book full of illustrations and explanations of all kinds of everyday objects as examined through a microscope.

Hooke had examined the point of a needle, fragments of linen and cork, and much more. The highlight of the book was a fold-out print of a flea, magnified to the size of a rabbit. The tiny creature had become a monstrous beast, with protective plates, legs covered with hair, strong claws, and two hanging sacs with stubbly hairs near its mouth.

The microscope offered the prospect of many other discoveries, Hooke wrote: there was a new world to explore and no detail so small that it could escape microscopic observation.[66] That prediction, and Hooke's observations, made his readers curious, and the *Micrographia* provoked them to conduct further research. And Antoni was one of them.

He visited England a few years after the book was published, after which he devised – as far as we know – the first research question he tried to solve using a microscope. It happened during a visit to the chalk cliffs near Chatham. They were so white that Antoni wondered how that was possible. He decided to examine the chalk through a microscope to find the answer.

He clearly had microscopes on his mind in those years. He closely examined an ant that had stung him, and found a small sting in its rear abdomen that is difficult to see with the naked eye. It is safe to assume that he examined it with a lens.[67]

Antoni travelled from the port of Harwich to the chalk cliffs, visiting London, the home of the Royal Society, on the way. Did he learn of the *Micrographia* there? Or was he already familiar with the book, and was that one of the reasons for his curiosity about England? Again, we have to guess, but for the reader who has had enough of all the uncertainties about Antoni's life, there is good news. No matter exactly what happened when and where, he had certainly discovered his love of microscopy and, within a few years, would be conducting groundbreaking research that would make him famous. From that time, we know much more for certain about his life.

Unlike Hooke, Antoni chose to use microscopes with a single lens – in fact, advanced magnifying glasses. In doing so, he trod in the footsteps of Amsterdam alderman and mathematician Johannes Hudde (1628–1704). From the late 1650s, Hudde had produced tiny lenses that, because of their strong curvature, enabled powerful magnification, enough for a microscope and with no need of a second lens.[68] That was an advantage, as a single lens produced less distortion than multiple lenses. Even though all microscopists covered the edges of their lenses, the light passing through the lens was still dispersed, though to a lesser extent.

Hooke was aware of that advantage and had also experimented with single-lens microscopes. But he found them unwieldy and impractical because, to get a clear image, you almost had to hold them in your eye.[69] Antoni was prepared to put up with the disadvantages, probably because his devices were more flexible than tube microscopes, which were usually mounted on a stand. He could move his microscopes any way he needed to and could allow the light to fall on his objects from different angles. That was to prove extremely useful. Working with these microscopes and interpreting the images they produced called for a certain dexterity, but Antoni possessed that skill.

Regnier de Graaf saw that and used his connections at the Royal Society to introduce his friend. In a letter to the Society's secretary Henry Oldenburg, he described Antoni as a 'most ingenious person' who made microscopes that far surpassed those made so far by others, and that enabled him to see more accurately than anyone else, and he was prepared to show the Royal Society samples of his work.[70]

The letter was accompanied by a description of Antoni's observations of objects that Hooke had discussed earlier in the *Micrographia*: fungi, bees and the body parts of a louse. With each of the objects, Antoni commented on the work conducted by Hooke. He asserted that a bee sting had not one barb, as Hooke had claimed, but two.[71] And the antennae of a louse was comprised of five parts, not four.[72]

It was a presumptuous entry into the prominent Royal Society, especially for someone completely unknown. But it worked. Oldenburg was enthusiastic about Antoni's work and published it in the journal *Philosophical Transactions*, which would some time later become the official journal of the Royal Society, and to which Antoni would write letters for fifty years.

That was difficult for him because of his insufficient knowledge of foreign languages. In those times, all self-respecting scholars had a command of French and Latin, and usually of Greek, but Antoni did not, or certainly only to a very limited extent. Constantijn Huygens Senior pointed this out when corresponding with the Royal Society about Antoni. He wrote that Antoni was 'of his own nature exceedingly curious and industrious', but was 'unlearned both in sciences and languages and specially, a philosopher who knows nothing of geometry'. Huygens's comment was written partly in Greek, to show that he was making an intellectual reference to classical antiquity, to the philosophical academy of Plato, to be precise, above the entrance of which were allegedly the following words, both a warning and an admonishment: 'Let no one ignorant of geometry enter here.' For Plato, mathematics, especially of the abstract kind, was a desirable and pure form of knowledge, and any profound thinker should have a good grounding in it.[73] That ideal still held sway in the seventeenth century, and Antoni did not fulfil it. He was good at arithmetic, but he had no training at the theoretical and philosophical level.

Moreover, to his regret, Antoni also spoke no English – which is striking given his many English connections and his past working in a branch that depended largely on English wool.[74] He wrote all his letters to London in Dutch, where they had to be translated. This language barrier created a tension that would determine Antoni's relationship with the external world for half a century. On the one hand, he was an innovative researcher, someone with interesting ideas and part of the international network of scholars that, around 1700, was so important in spreading knowledge, and from 1680 he was even a fellow of the Royal Society. That was the side of Antoni that had been immortalized by Verkolje. But on the other hand, he was a man with limited knowledge of the so-important classics and their languages, and with no academic background.

Constantijn Huygens Junior made light of this tension in a letter to his brother, Christiaan. In gossip mode, he wrote that everyone honoured Antoni as the greatest man of the century. According to Constantijn Junior, that had gone a little to his head: Antoni had asked Constantijn Senior whether his status as fellow of the Royal Society made him the social equal of a trained doctor. 'It has made him a little conceited,' he wrote.[75]

Antoni had risen to great heights, but his lack of a university education would never be forgotten.

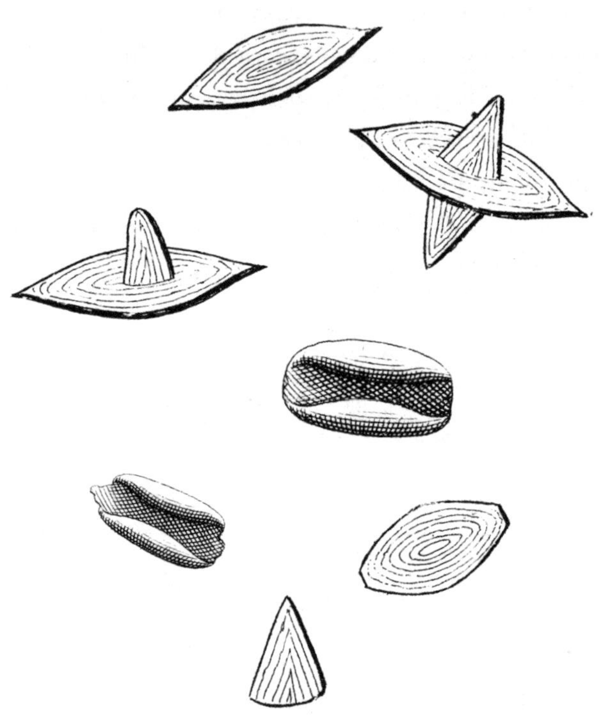

THREE

Globules and Stalks

While Antoni was doing his work at the town hall and exploring the possibilities of lenses, ships from the Dutch East India Company (VOC) were sailing to Asia to purchase luxury goods. It was a brutal form of trade that entailed the use of much force and violence on the part of the VOC. But, as with the majority of his contemporaries, this was of no concern to Antoni. He was, however, interested in the exotic products that the VOC brought to Delft, and in pepper especially.

How did pepper get its taste? And why was it so spicy? In 1676, Antoni came up with an explanation. He had examined peppercorns under the microscopes he had now been using for a few years. He discovered that each peppercorn consisted of more than 50,000 'rather long' particles, blunt at one end and pointed at the other. Perhaps, he reasoned, the sharp ends of these small stalks pricked your tongue as you ate them and that caused the 'heat of the pepper'.[1]

With this theory, Antoni showed himself a thinker of his times, as many researchers were convinced that minuscule particles, like the tiny stalks in pepper, could explain all natural phenomena, from the taste of spices to the orbits of planets. The most prominent proponent of this idea was René Descartes (1596–1650). This mathematician and philosopher had announced that all things, living beings and other physical objects, were made up of small particles that were invisible to the naked eye. According to Descartes, these particles were all closely packed together, with no space between them, and filled the entire cosmos. The particles had a physical form and they moved, but otherwise they had no properties – no colour, no scent or anything else. Their 'behaviour', and that of the larger objects and beings that they made up, was determined by their form and motion. So if you understood the particles, you could understand nature itself.

In the seventeenth century, that was a progressive idea. Descartes and like-minded contemporaries explicitly rejected the way of theorizing taught at many universities, which was strongly influenced by the ancient Greek philosopher Aristotle. In the fourth century BC, Aristotle had taught that all things, plants, animals and people had their own 'form'. By this, he did not mean a physical form, such as round or square, but something much more abstract that made a tree a tree, a human a human and a peppercorn a peppercorn. According to Aristotle, that form determined why things did what they did: why a tree had leaves and why pepper tasted hot.

For seventeenth-century particle theorists, that sounded good but explained very little. They were not interested in abstract forms but in mechanisms: how did the small particles in the natural world interact with each other and what outcome did that have in the visible world? How did it result in plants, animals, humans and celestial bodies?

In the Dutch Republic, this alternative perspective became popular among groups of doctors. They had been trained in humour theory – based on the four kinds of bodily fluids – which was closely related to Aristotle's view of nature. But, under the influence of men like Descartes, they started looking for other

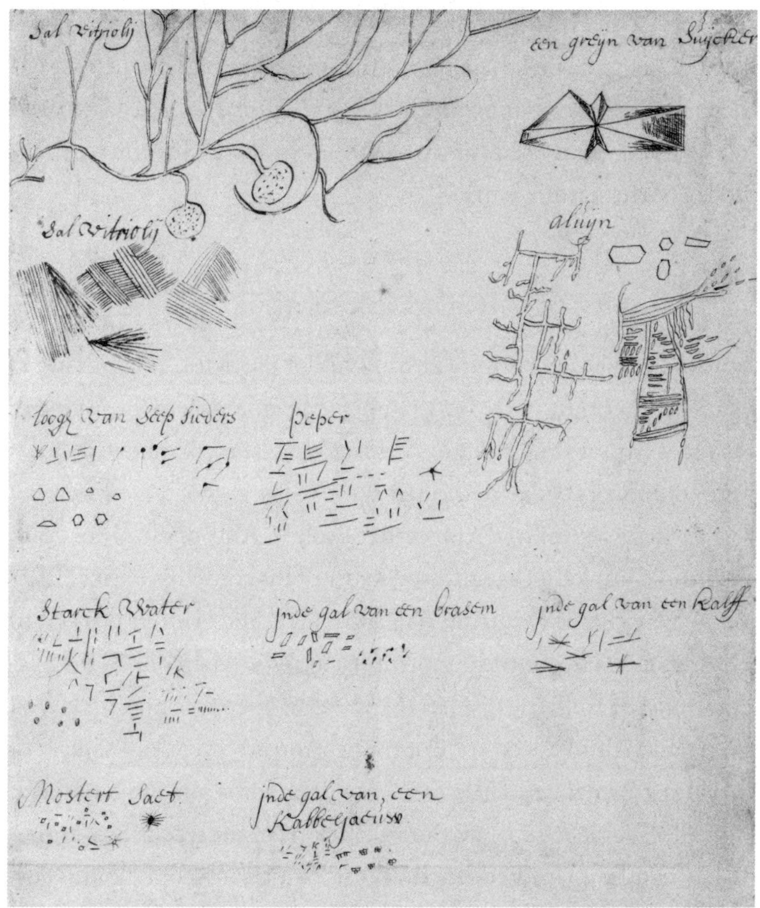

14 Antoni's portrayal of sharp pepper particles, in the centre of the page, and the structures of other food products with a clearly defined taste, including sugar (above right) and mustard seeds (below left).

explanations for sickness and health. They exchanged the theory of the body as a container filled with fluids for a mechanistic view of the body as a machine, in which each organ had its own purpose.

Regnier de Graaf's mentor in Leiden, Sylvius, was one of them. He combined his iatrochemical way of thinking with a mechanistic one and passed his ideas on to students like De Graaf.[2] The latter may then have passed them on to Antoni, or he may have discovered them elsewhere, but one thing is certain: Antoni became fascinated by the theory and used his microscopes to search for the building blocks of the natural world.[3]

Remarkable forms

Antoni's earlier letters were already full of particles, or 'globules' as he called them. In 1674, for example, he explained to Constantijn Huygens Senior how hair grew. He had observed that new globules were continually added to the hair at the roots, so that it grew longer.[4]

Particles also played a prominent role in Antoni's study of gout, a painful complaint that affected many of his contemporaries. Their joints were full of a calcium-like substance, and Antoni wanted to investigate it. He got the opportunity in 1679, when one of his relations suffered from the disease: there was an open sore near his elbow, from which fluid oozed continuously.[5] Antoni collected a sample of the fluid in two unused glasses. It contained some pus and blood, but Antoni was able to separate the 'calcium' from the rest using a knife.[6]

When he examined the fluid under a microscope he was surprised, as it did not look like the limestone he had studied earlier. It was made up of minuscule and very thin, transparent shapes, pointed at both ends. They were incredibly strong, leading Antoni to conclude that the particles comprising them must be very closely packed together.[7]

According to him, that also explained the pain, because the sharp particles pricked the patient, causing an unpleasant sensation.[8] Fortunately, the affected parts of their bodies gradually became numb and they did not feel the pain continuously. It only flared up if the patient banged the swollen joint against something or if it suffered vigorous contact in some other way.[9]

*

Antoni also saw interesting particles outside the human body: in animals and plants, and especially in inorganic objects and substances. He was particularly fascinated by salts, and he extracted them from bladder stones, coral, medicines he bought from an apothecary, material he picked out from between his own toes and so on.[10] The particles had wonderful shapes: elongated or hexagonal, flat or with branches like a tree – anything you could imagine.

He also observed interesting shapes in stones. For example, during a visit to the city of Utrecht, he examined the remains of the nave of the Dom Church, which had been destroyed during a summer storm in 1674. He became fascinated by a red stone used in pillars and tombstones, and took a piece home to study more closely. The red colouring proved to come from a substance that was of no further interest to him. It was the rest of the stone that caught his attention. It consisted of grains of sand with sharp points, at least as beautiful as a cut diamond. And those grains proved to be made up of even smaller grains, with equally beautiful shapes.[11]

These kinds of observations confirmed an idea that Antoni had been thinking about for some time: that the small particles he observed were made up of smaller particles, and they in turn of even smaller ones, and so on ad infinitum.

Snail shells

For more than fifty years, Antoni would search for globules, and for the particles of which they were composed. He wanted to know what they looked like, how they moved and how they became part of bigger wholes (people, animals and so on). After all, there lay the root cause of all change, and thus of everything that happened at the macro level.

In 1704, Antoni applied this approach to the study of a curious phenomenon from the Swiss Alps. He had heard from an acquaintance that an abnormal kind of snail shell was to be found there. The shells seemed to be made of different material than regular snail shells and were much heavier.[12] We now know that they were the fossilized shells of squid-like creatures called ammonites that swam in sea water in the distant past. But that was still unknown in Antoni's time, so he had to research the strange shells himself. The acquaintance gave him a few to examine, and Antoni broke one into pieces. He observed a metal-like substance on the fragments, which he had seen before with objects that contained a lot of sulphur. So he heated a fragment of shell and, indeed, sulphur was released.[13]

Antoni explained all this by concluding that, where the snails had died, the ground must have had a high sulphur content. The sulphur gradually seeped out of the ground and was absorbed by the shells, through pores that Antoni claimed to have seen in their outer skin. That made the shell heavier and gave it its metallic appearance.

Newton's weakness

As Antoni's study of the snails shows, he also conducted research without microscopes. Another example of this was when he acquainted himself with the ideas of Isaac Newton (1643–1727). In 1687, Newton had published a book entitled *Philosophiae Naturalis Principia Mathematica*, which contained laws that explained, among other things, the motion of the planets. Newton received great acclaim for the book, but there was a serious problem with his theory. One of its central principles was that of gravity, which could work over enormous distances. That was a tricky issue for particle theorists, as everything in their model was based on direct contact between colliding particles. They did not accept the notion that planets, for example, could attract each other without touching. Consequently, they considered the as yet unexplained force of gravity as a major weakness in Newton's work.

Christiaan Huygens also found it difficult to accept and so he published an alternative theory.[14] According to Huygens, though it looked as though gravity was active from a distance, there were actually minuscule particles at work that, for example, pushed heavy objects towards the Earth.[15] This enabled him to accept Newton's clever calculations and still hold on to his theory of particles.

Antoni looked at the question in a similar way. It was not that he drew up a physical theory about it; he was not the man to do that. But, in 1696, he did describe his demonstration model of the rotating Earth, with particles that kept everything in its right place. The model consisted of a bulbous bottle that he filled with water to represent the atmosphere. Inside the bottle, he suspended a lead ball, which represented Earth. In the water, he scattered fragments of red pigment, which he had pulverized with a hammer. These stood for the clouds in the sky.[16]

If he rotated the bottle, everything looked like it does in reality, with Earth (the lead ball) in the middle, the air (the water) all around it and the clouds (the fragments of pigment) high up in the atmosphere. That happened, according to Antoni, because the particles pushed each other into their correct places, without the need for an incomprehensible concept like 'gravity'.[17]

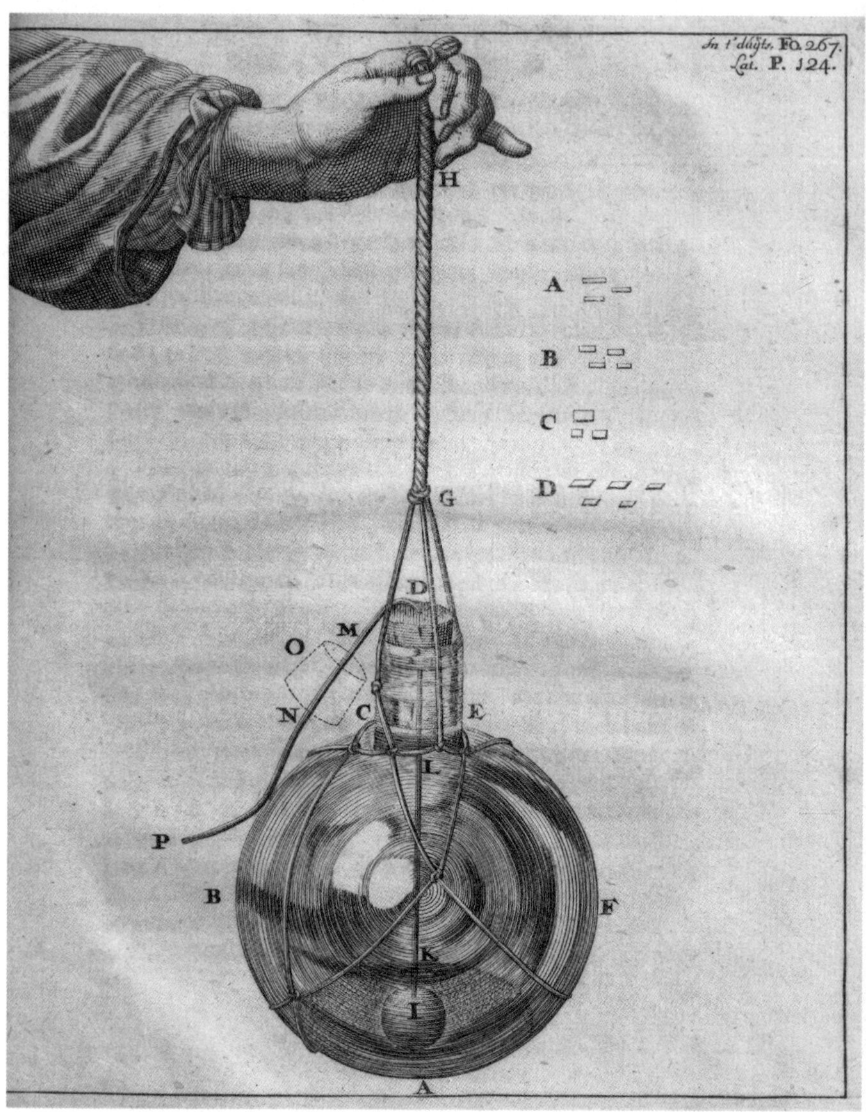

15 Antoni made a model of the rotating Earth, to demonstrate how moving particles were responsible for the composition of the atmosphere.

With hindsight, Antoni was wrong and Newton right. But Antoni's attempt to capture Earth and its atmosphere within a particle model shows just how broad his interests were and how strong his faith in particle theory.

*

Much earlier, in 1673, Antoni had already studied small particles in the air without the use of a microscope.[18] To do that, he made a sort of syringe, but without a hole in the end. He took a thin glass tube that was closed at one end and half filled it with water. He then took a piece of rigid copper wire with a leather washer around one end and pushed it into the glass tube, like the plunger in a syringe. To make sure that no water or air could leak out of the tube, he had soaked the washer in candlewax.[19] So now he had a tube with a little water and air in it, and a plunger that he could move in and out. If he pushed the plunger inwards, the space in the tube became smaller. But how could that happen if the space in the tube was full with air and water? For Antoni, there was only one explanation possible: minuscule air particles were escaping from the tube through tiny pores in the glass – so small that he could not even see them with his lenses.[20]

For modern readers, this may seem like a strange way of reasoning, as they know that air cannot pass through intact glass. But for Antoni, it sounded logical and, on this point, he was following Descartes. After all, the latter had described a world full of particles, with no empty space between them. That meant that one particle could only move if another made way for it. So the particles of the plunger could only move into the tube if other particles left it. And, as there were no visible openings in the glass, that had to occur via tiny invisible holes.

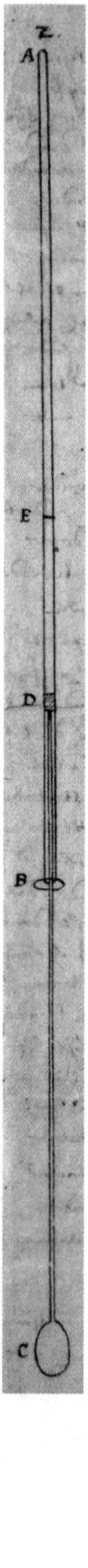

Thus far, Descartes would have agreed with this explanation, if he had still been alive when Antoni recorded it. But then it took an unexpected turn. According to Antoni, the smallest air particles were the same size as grains of sand, with a slightly larger category ranging from a large grain of sand to a redcurrant, and a third category sized between a redcurrant and a cherry.[21] So, for tiny particles, they were quite a size. Yet, Antoni still maintained that they could pass through minuscule pores, because they had 'a soft and fluid body'.[22] So they could flow though invisible openings like small sacs of water.

These claims would have surprised Descartes. His particles were not at all soft and fluid, and were a lot smaller than grains of sand. Antoni had clearly taken some of Descartes' philosophy on board, but had not applied it to the letter.

The reader may wonder why Antoni was so fascinated by air particles. To answer that, we need to go back to his predecessor Robert Hooke. Hooke had gained a lot of research experience in his younger years, working with Robert Boyle. And Boyle, also a proponent of particle theory, was interested in air and its effects on its surroundings. He had Hooke construct an air pump, which they used to conduct a whole range of experiments that Boyle described in 1660 in his book *New Experiments Physico-Mechanicall, Touching the Spring of the Air and Its Effects*. In the book, Boyle claimed that air had elastic properties – the 'Spring' in the title. To explain what

16 For his research into the composition of air, Antoni used a glass tube that was closed at one end (at the top in the figure). He inserted a plunger (at the bottom), which could close the tube off completely.

he meant by that, he compared air with a fleece of wool, composed of a large number of fine hairs. Normally, the hairs take up a lot of space, but if you applied pressure to them, they would compress into a thin layer. When you released the pressure, the wool would expand again. According to Boyle, air behaved in a similar way: it would contract or expand depending on the pressure exerted upon it.[23]

Boyle would thus have explained Antoni's experiment as the plunger pushing the air particles closer together, just like the hairs in the compressed wool. How that worked exactly at micro level, Boyle was unable to say, because the particles were too small to observe. Perhaps they really did look like fine hairs with empty space in between, or perhaps they looked completely different. Experiments gave no conclusive results and Boyle wished to keep speculation to a minimum. That's why he did not go into detail.[24] But he did not agree with Descartes that the whole cosmos was completely filled with particles.

While Boyle thought that the particles were too small to see, his assistant Hooke was optimistic about observing them. So, after doing the air-pump experiments with Boyle, he turned to the microscope. And that hope is clearly expressed at the start of his *Micrographia*: 'And by the help of microscopes, there is nothing so small as to escape our inquiry.'[25]

That Antoni began to talk about air pressure in his second letter shows that he was aware of the background of Hooke's work. Remarkable, considering that Hooke and Boyle published in English and Latin, languages that Antoni by his own admission could not speak. And there were as yet no translations of their work. So how did he know about their research? More learned acquaintances like De Graaf and, in later years, Christiaan Huygens – also a particle theorist – will certainly have played a role.

A little English?

In addition, readers of Antoni's letters can get the impression that he occasionally read parts of books and articles, even though they were usually written in languages that he claimed not to understand.

In a letter from 1675, for example, he responds to an article in *Philosophical Transactions*.[26] The article described how Robert Boyle had experimented with hollow glass tubes. He hung coins on the tubes and suspended them in liquid to find out how much they weighed. It was an accurate method of measurement: the heavier the coin, the further the tubes sank into the liquid.[27]

Boyle had described the experiment in pages full of text in English. Antoni wrote that the text was too difficult for him to follow, but he was able to decipher what Boyle was saying from an accompanying illustration. That was very informative for him, as he had also conducted experiments with glass tubes. That knowledge enabled him to comprehend what Boyle had done without understanding a single word of the article.[28]

That was quite an achievement, as the illustration was not very detailed. It did not, for example, show the liquid in which the glass tubes were to be suspended. Moreover, Antoni had conducted different experiments with his tubes than that described in the illustration: he had not hung coins under the tubes, but had used the latter to determine the viscosity of the liquids: the higher the viscosity, the less far the tubes sank into the liquid.[29]

And that's where it gets complicated. Boyle had also used the glass tubes for the same viscosity tests as Antoni, but had not included an illustration of the tests in his article.[30] The tests are described only in the English text, which Antoni claimed not to

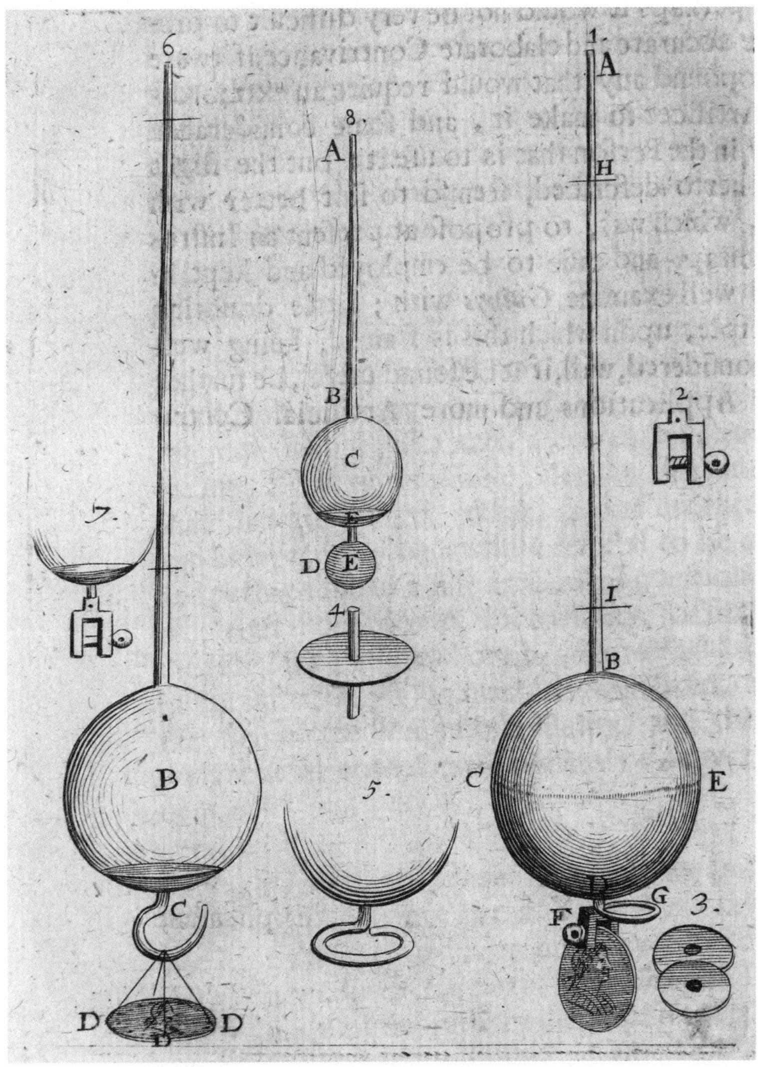

17 Robert Boyle hung coins on hollow glass tubes and suspended them in liquids to determine the weight of the coins.

be able to understand. This suggests that he may after all have been able to read a little English.

There are more indications that this might have been the case. Let's go back to Hooke and his *Micrographia*. In the second chapter of this book I explained that, in his first letter, Antoni had a number of critical comments on Hooke's observations: his descriptions of a bee's stinger, the structure of a louse and the mould growing on leather. All three descriptions were accompanied by illustrations, but some prior knowledge was required to recognize them, especially the illustration of the mould.[31]

18 What Robert Hooke saw after he removed hairy mould from a leather book cover and placed it under a microscope.

In theory it is possible that, separately from Hooke, Antoni scraped the same mould from leather, had seen the same growth and could therefore understand the illustration. But that's not what happened. According to Antoni, the mould looked different from how Hooke had described it. The illustration in the *Micrographia* showed flower-like structures but, according to Antoni, the mould consisted of thin transparent stalks topped off with bulbous growths that were a hundred times thicker than the stalks. Those growths consisted of all kinds of smaller bulbs.[32]

It is starting to look very likely that Antoni did understand some of what was described in the English texts: either by puzzling it out using what English vocabulary he had picked up, perhaps when he was working with Davidson, while travelling in England, or with someone else's help.[33]

A further pointer in that direction is to be found in his eighth letter, also addressed to the Royal Society.[34] Antoni writes that he had examined cork, the pith of elder wood and the white (centre) of a quill. He did not explain exactly what he had seen but did enclose thinly cut pieces of these items along with the letter, so that his readers in London could examine them themselves under a microscope.[35]

Why did Antoni explicitly choose cork, elder and a quill? The *Micrographia* offers a clue. Hooke describes the airy structure of cork and notes that he had observed a similar structure in the core of all plants, including the elder, and in the core of birds feathers.[36] There were no illustrations accompanying these descriptions, reinforcing the suspicion that Antoni knew what was in Hooke's text, and that he could have derived his knowledge of particle theory and microscopic research, to a limited extent, from English sources.[37]

Red globules

All in all, it is therefore quite feasible that the *Micrographia* was a source of inspiration to Antoni. But he went much further than the book and studied a multitude of subjects that Hooke had passed by. To do that, he even used his own body, for example, by tapping off blood. When he examined that through his lens, he saw a relatively clear liquid containing red globules – what we now call red blood cells.[38]

Antoni was incidentally not the first to see the little red globules. The Italian Marcello Malpighi (1628–1694) had already observed them in 1666, but Antoni did not know that.[39] And another Dutch microscopist, Jan Swammerdam (1637–1680), may also have seen them. Like De Graaf, Swammerdam was a pupil of Sylvius and a prominent researcher of the human body.[40] He observed in blood an 'infinite number of round particles'.[41] But it is not clear exactly when that happened, as Swammerdam recorded his observations in a book that he never published himself and that only appeared decades after his death.

Antoni studied the globules in much greater detail than Malpighi and Swammerdam. It gave him 'the greatest of pleasure' to see them moving around. One stormy day, he took a sample of blood outside and noted that the red particles moved in tune with every gust of wind.[42] This cannot be true, as is evident to us today, as the blood was in a thin glass tube. The effect was probably caused indirectly by the wind moving Antoni's hand, and thereby the tube. But Antoni's great pleasure was genuine and would drive his research for decades: he would never tire of frolicking particles, and he continued to look at them through his microscope, year after year after year.

The pleasure aside, Antoni had scientific reasons for continuing his observations. One observation could rest on coincidence, and examining a single subject – his own blood – only provided limited information. So after looking at the blood from his own hand, he began to wonder if other living beings had the same globules running through their veins.[43] That question was the start of years of research into the blood of a wide variety of creatures: oxen, sheep, rabbits, salmon, cod, frogs and more.[44]

He could purchase much of his research material close to home as, on the other side of the canal to his house was a fish market and, immediately behind that, a meat market. Merchants in the city supplied him with all kinds of animals and interesting information. In 1677, for example, a fish merchant told Antoni that eels' blood was dangerous. If you got it in your eye, you would suffer 'incredibly deathly' pain all day.[45]

19 The fish and meat markets opposite Antoni's house were sources of information and research materials.

That made Antoni curious, and he immediately bought six eels to examine their blood. It proved to contain 'small rods', twice as long as human blood globules, and a third of their thickness.[46] The ends of the rods were sharp, explaining the painful effect on the eyes, just as the pointed particles in pepper were responsible for its spicy taste.[47] Perhaps, Antoni wrote to secretary of the Royal Society Henry Oldenburg, the rods were the result of blood globules drying out. That made them hard, so that they pricked so painfully.[48]

The explanation was not very satisfactory, as Antoni did not make clear why eels' globules formed the painful rods while human globules, for example, did not. That made Antoni himself doubt his theory and he conducted further experiments.[49] He discovered that there was a difference between the globules of mammals and those of aquatic animals and birds: the globules in the former were spherical and in the latter flat and oval.[50] These were at least the basic shapes as, under pressure, the globules seemed to change their appearance.[51]

Readers with some biological knowledge may now be confused, as they know that all red blood cells are disc-shaped. They also have an indentation in the centre, making them look a little like doughnuts, with a small layer in the hole. Antoni saw this, too, in the blood of a salmon.[52] But he would not believe that human blood also contained discs. Not even when he saw them with his own eyes.[53]

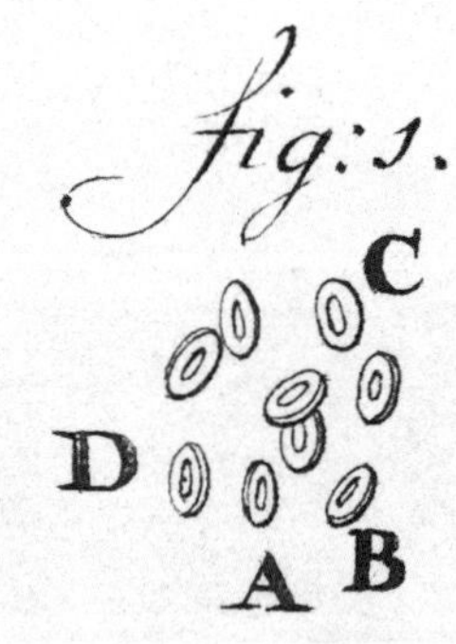

20 Antoni observed doughnut-shaped particles in the blood of a salmon. We now call them red blood cells.

That happened in 1716, when Antoni had once again pricked his thumb with a needle to obtain some blood, which he mixed with water. That caused the red globules to separate so that he could examine them individually.[54] To his amazement, he saw that the red particles were quite flat, with an indentation in the centre – just as with the salmon. He decided that this was not their natural shape, and was caused by their flexibility. When placed on a flat glass surface, they would collapse slightly.[55]

Everything in sixes

After examining them for a long time, Antoni concluded that all blood globules – spherical and flat – consisted of six smaller globules. He suspected that each of these six were also made up of six even smaller ones.[56] Present-day biologists see no evidence at all that this is true, begging the question of why Antoni thought he saw this structure based on clusters of six globules. We can

21 Antoni made wax models to show that (more or less) spherical blood globules could be made up of six smaller globules, and the latter of six even smaller ones. He also believed that flat, oval globules consisted of six smaller particles.

speculate that, during his observations, his blood cells shrivelled up and gave the impression of consisting of multiple particles. Or that they burst when immersed in water, so that they looked different than normal. This hypothesis has been tested by researchers from Visualizing the Unknown, which replicates the work of Antoni and others. They examined blood cells soaked in water and dried up samples, using Antoni's original lenses. But the results were disappointing: the molested blood cells did not look as Antoni had described, and they saw no evidence at all of structures based on clusters of six.

But Antoni's idea that blood globules consisted of smaller, similar particles fitted in well with his version of particle theory, in which all the building blocks of nature consisted of progressively smaller blocks and biological processes came down to rearrangements of the basic particles.

One important biological process started when someone ate food, breaking it up into smaller pieces in the mouth and intestines. Because the stomach and intestines continually kneaded the food, Antoni thought, it was transformed into a watery substance.[57] That was absorbed by blood vessels, in which it coagulated to form mini-globules. These mini-globules would eventually cluster in groups of six to make the red blood globules.[58]

Once the blood globules were fully formed, they could perform other functions. That required small changes of appearance. They could, for example, turn green to make the 'phlegm' that came up out of Antoni's throat if he did not feel well.[59] Or they would turn white, and form the basic particles of milk.[60] Here Antoni was pursuing the idea that De Graaf had tried to prove with his failed transfusion experiment that blood and milk were fundamentally the same.

These ideas were attractive in their mechanistic simplicity.[61] And they seemed to relate to universal principles, as Antoni also believed that he could observe particles made up of six sub-particles outside the human body.[62] In fermenting beer, for example, he claimed that grain particles broke up and then coagulated together again to form yeast particles, which looked strikingly like red blood globules: they were the same size and were made up of six sub-particles.[63] Antoni also found the same kind of globules, made up of six particles, in the dregs of his wine, but they were a little smaller.[64] Just to be clear, anyone who examines blood, milk, beer or any of the other fluids Antoni studied through a microscope will see no sign of the sixfold pattern. But Antoni was convinced that it existed.

Why the particles were structured in this way, he was unable to say. He suspected that, during Creation, God had determined that it should be so and ensured that blood particles fitted nicely into veins.[65] He also had nothing conclusive to say about the exact nature of the relationship between the wine-dreg, yeast and blood particles. He suspected that they were connected, but precisely how he didn't know, just as he was unable to clarify how blood particles changed into mucus or milk particles.

No transformations

Here lay the limits of the explanatory capacity of Antoni's ideas. He could observe globules, rods, discs and many other shapes, and speculate how they moved. But that was as far as he could go. In his mechanistic world, chemical processes were irrelevant; the notion that substances could react with each other and consequently acquire fundamentally different properties was alien to him. That made things difficult, as living organisms are in reality bursting at the seams with chemical processes.

It is somewhat unfair to blame Antoni for not taking account of chemistry, as it would not fully develop as a science until the late eighteenth century, decades after his death. Consequently, he lacked all kinds of knowledge that would have made his life a little easier – that there was a substance called 'oxygen', for example, and that it was crucial for life, and that red blood cells carried it to every remote corner of our bodies. Because Antoni did not know that, he had to guess at the purpose of the red globules. He wondered whether they served as fuel, but jettisoned the idea after studying the circulation of blood. The globules remained permanently in the blood and were not used up, as happens with food.

At the same time, he had noticed that part of the blood – which he called serum – could escape from the veins, in places where they were small and very thin. That led him to conclude that the serum was the fuel.[66] So what did the globules do? That remained, logically speaking, unclear.[67]

Antoni was restricted by the simple fact that he was exploring unknown territory, and could not make use of the work that later researchers added to his observations. But sometimes it was more than that, and his particle theory sent him off in the wrong direction – though that is easy to say with the wisdom of hindsight.

Take, for example, his study of the cause of gout. In his time, it was widely assumed that this painful complaint was caused by too much wine and sex. But Antoni thought that couldn't possibly be true.[68] He considered the sex explanation not even worth refuting, as it did not fit in logically with his particle theory. He took the wine hypothesis a little more seriously, however, and explored it in his own typical way, comparing the crystals in wine with those in gout swellings. He reasoned that, if they had a similar shape, the gout rods could come from wine. But the wine crystals looked completely different to those from gout sufferers, so Antoni concluded that there was no link between the drink and the disease.[69]

In hindsight, that was an overhasty conclusion. Today, doctors tell patients to avoid alcohol because wine and especially beer raise the risk of gout attacks.[70] As it would have been impossible for Antoni to discover the complex mechanism behind that connection, he should not be blamed for failing to see it. But his resolute conviction that particles and substances never really change led him to reject a useful theory all too readily.

In other words, Antoni's mechanistic particle theory encouraged him to spend fifty years examining the world around him and to discover as yet unknown minuscule creatures and particles: the theory was the intellectual stimulus behind his observations. But at the same time, his view of the world limited his understanding of the things that he saw.

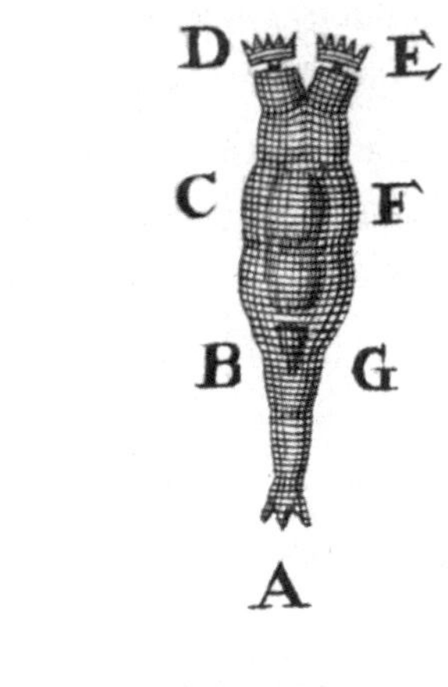
D
E
C
F
B
G
A

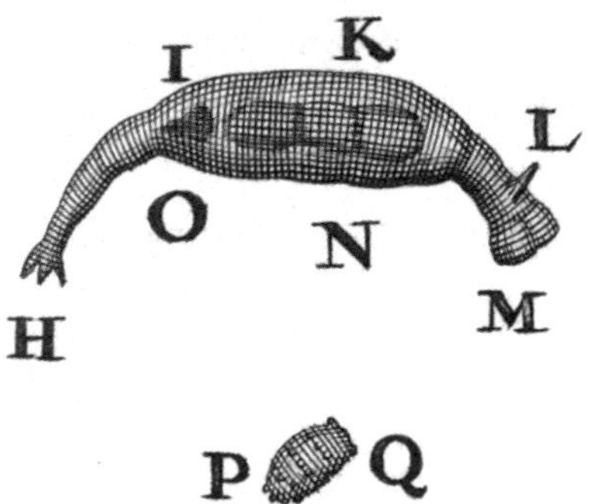
I
K
L
O
N
H
M
P
Q

FOUR

Mind-Bogglingly Multitudinous

One day, Antoni crossed the Markt in Delft with a metal instrument under his arm in the shape of a semi-circle.[1] It was a quadrant, which is used to measure angles, and Antoni pointed it at the tower of the Nieuwe Kerk. An acquaintance of Antoni's called Jacob Spoors (*c.* 1595–1677) did the same. They wanted to know how high the tower was, and the quadrant was the perfect instrument with which to find that out. Antoni and Spoors each took a measurement, so that they could compare each other's results.[2]

They started by making a rough estimate of the height of the church. Then they took up positions about the same distance from the foot of the tower as the height they had estimated. They drew an imaginary line diagonally upwards from their eye to the top of the tower, and then another horizontally from their eye to the tower. Holding up the quadrant in front of their faces, they could determine the angle between the two lines. If the angle was 45 degrees, they were in the right place and the distance to the tower would be the same as its height. They had to take account of the fact that they were probably holding the quadrant a few feet above the ground, and had to add that to their result. If the angle shown by the quadrant was larger or smaller than 45 degrees, they had to move forwards or backwards until they were in the right place.[3]

This was all explained in detail in books for land surveyors. The surveyors were specialists who mapped towns, cities, fields and the rest of the world. That meant they had to be good at geometry and possess all kinds of skill to determine distances, curvatures and angles in the landscape. Antoni was familiar with these methods, as he had taken a land-surveying examination and was 'found to be proficient'.[4] Once he had the required knowledge and skills, he could work as a paid land surveyor.

In the late 1660s, besides his first microscopic observations and his bookkeeping work at the town hall, Antoni was thus devoting his attention to something else new. His interest in land surveying was, however, not entirely unknown terrain for him, as it harked back to the mathematical work he undertook during his time with William Davidson in Amsterdam, and later when checking the financial accounts of Sijmon Bourbon.

Moreover, his surveying skills were also valuable at the town hall, where Antoni was already working. The city councillors wanted to

22 A land surveyor measuring the height of a tower.

know how much land the townspeople owned so that they knew how much to levy in taxes.[5] As a trading centre, Delft also needed officials who were able to bring some order to the bewildering tangle of measurements and weights used by merchants in the seventeenth century. They had to determine the quantity of goods being traded in the town, again so that they could levy taxes on them. Wine, for example, was transported in casks. To keep the administration manageable, the casks were supposed to have standard volumes. To check that, Delft appointed a *wijnroeier*, a wine gauger who measured the volume of all the casks and gave them a stamp of approval. It was not difficult work, but the bulbous shape of the cask called for some geometric knowledge. As a trained land surveyor, Antoni was appointed assistant wine gauger in 1670, and city wine gauger in 1679.[6]

Another surveying project conducted at that time was the *Kaart Figuratief*, a detailed map of Delft, which was published in 1678 and fragments of which have appeared earlier in this book. Spoors, also a land surveyor, performed the measurements for the map and it was logical in this capacity that he and Antoni measured the height of the Nieuwe Kerk.[7]

*

The two men may have met earlier, as they had many things in common. Like Antoni, Spoors was a man who worked with files and administration, as a solicitor among other things. He was also a member of the Sint Nicolaas Gilde, the merchants' association that Antoni had joined when he set up his textile shop.[8] And in 1659, Spoors's son married one of Antoni's nieces.[9]

Moreover, Spoors conducted experiments with sight and lenses, though he was more interested in looking into the far distance. In 1633, when Antoni was still a baby, he had placed a burning candle on top of a dyke to determine how far the light reached.[10] He walked 800 *Rynlansche roeden* – some 3 kilometres (2 mi.) – from the candle and could still see the flame, like a star in the sky.[11]

Shortly afterwards, he got hold of a telescope and looked at the sky, the sun in particular.[12] In 1638, he wrote a book in which he argued that the sun, 'that most brilliant of Creations', was the centre of the cosmos, just as Antoni would almost sixty years later, using his model of a rotating bottle filled with water and fragments of pigment.[13]

A thousand times smaller

Like Regnier de Graaf, Spoors could have been an influence on Antoni. He may have done that in a practical way, by sending Antoni to the place where he would make his most sensational discovery. In the summer of 1674, Antoni visited the Berkelse Meer, two small lakes to the southeast of Delft.[14] The water looked strange – white, with patches of green – so Antoni put a sample in a bottle to look at through his microscope later. What he saw was 'wondrous to behold': little particles, joined together by tiny green strands curled in spirals, and minuscule creatures, roundish, oval and elongated, in all kinds of colours. They were all racing around in all directions – up, down and round and round. Everything was unimaginably small: more than a thousand times smaller than the smallest animals he had seen until then 'in the rind of cheese, wheaten flour, mould, and the like'.[15]

We now know that what Antoni had seen were in fact micro-organisms, including single-cell life forms, and he is still celebrated for this discovery. We will return to this breakthrough later, but now the question is, what did he do there, by those lakes?

One clue is to be found on a map that Jacob Spoors published a year later, of a piece of land near the 'Oostermeer at Berckel'.[16] The two lakes at the Berkelse Meer were joined by a small water course. Spoors had surveyed the eastern lake, because a man called Adriaen Coornwinder had plans to construct a new drainage system there so that the lake could be reclaimed.[17] To do that, everything had to be accurately measured and mapped out.

It is questionable whether Spoors did all the surveying work himself, as he was by that time around eighty years old and the land around the lake was boggy and difficult to move about on.[18] Perhaps he was assisted by a younger colleague, such as Antoni, for example. A few years later, in 1677, Antoni bought a plot of land near the Berkelse Meer from Adriaen Coornwinder's widow. He may thus have met her when he was helping Spoors survey their land.[19]

Although this is all speculation, in a letter describing his observations, Antoni certainly described all the tiny strands and other animals as a real land surveyor. Fascinated by their minuscule size, he set about calculating how big they were compared to a grain of sand. In effect, he was using the same trick as present-day journalists who compare large areas with football fields or crowds of people with full stadiums: he brought the unimaginable down to the level of the everyday. Grains of sand were already a tried-and-tested subject for comparison, as Greek mathematician and thinker Archimedes had used them to make calculations two millennia earlier. In his book *Psammites* (The Sand Reckoner), he had tried to determine how many grains of sand fitted into the universe. It is possible that Antoni was aware of Archimedes' calculations since, as a land surveyor, he knew of the latter's geometric insights.[20] In any case, his own calculations were in the opposite direction: rather than working out how many grains of sand fitted into the universe, he determined how many of his little animals fitted into a grain of sand. He used that yardstick for many years, until he decided in his old age that grains of sand varied too much in size. He then switched to grains of barley or mustard seeds, which were much more consistent.[21]

After considerable calculation work, he concluded that he could lay three to four hundred of the animals end-to-end along the length of a single grain. To avoid his readers thinking that he was exaggerating, he chose the lowest number – three hundred – to calculate further. Assuming that the volume of a grain of sand was equal to its length × length × length – as with a cube – he concluded that 300 × 300 × 300 (that is, 27,000,000) little animals fitted into one grain.[22]

That was in itself an impressive number, but to make it even clearer to his readers just how inconceivably small the animals were, Antoni continued with his calculations, using the even more familiar unit of a *duim* (thumb), the equivalent of an English inch. According to him, there were 80 grains of sand to an inch, and so 27,000,000 × 80 × 80 × 80 (that is, 13,824,000,000,000) creatures in a cubic inch. In other words, more than 10 trillion.[23]

Trillions of blood vessels

According to Antoni, the minuscule size of the animals was even more mind-boggling. To fully understand what he had to say about that, a few words of explanation are in order. In 1628, British doctor William Harvey published *De Motu Cordis*, in which he argued that blood circulated around the human body. Doctors had long been taught that blood flowed from the heart to the rest of the body, where it was used up so that new blood had to be continually produced. But Harvey used tests to show that this was incorrect and that blood indeed circulated.

That meant that there had to be a (more or less) closed system through which blood flowed from the heart to the other parts of the body and back again. We now know what that system looks like, and that minuscule capillaries run between large arteries on one side and veins on the other side. But Harvey was writing well before the heyday of microscopic research and was unable to see the capillaries. He could only assume that such vessels connecting arteries and veins existed.

Antoni did see the capillaries, though – as with red blood cells – he was not the first. Malpighi and Swammerdam had once again been ahead of him, but here too he described his discovery in greater detail. In doing so, he provided conclusive proof that blood circulated through the body – though, as he himself noted, some doubters would continue to insist years later that arteries had open ends through which the blood flowed and was used up.[24]

The first time he discussed the capillaries in detail was in 1679, in the letter in which he also presented the calculations on the minuscule animals and grains of sand described above.[25] He didn't talk about their biological importance, as he was then more concerned with their 'incomprehensible smallness'.[26] And he was especially fascinated by the even smaller capillaries that he believed must be found in the microscopic animals.[27]

Antoni assumed that the smallest of creatures were made up of the same fundamental particles as larger animals, and thus contained similar structures: heads, legs, fins and so on. He even thought he saw these features when he examined the life in the water of the Berkelse Meer.[28] He was incorrect, as we now know with hindsight, but the idea fitted in well with his particle theory, in which fundamental building blocks behaved in the same way.

That's why Antoni thought it logical that even the smallest of animals had blood vessels.[29] He also assumed that all animals had similar proportions, that their various body parts had the same relative sizes.[30] So a creature that was a thousand times smaller than a human would have eyes, legs and capillaries that were a thousand times smaller.

To show clearly how astoundingly tiny the capillaries of his little animals were, he made a new calculation. But this time, he took a somewhat strange route and ultimately a wrong turning. First, he determined how many times the thickness of a human capillary fitted into a hair from his wig (33 times); then how many times the hair fitted into an inch (600) and how many inches the diameter of his waist measured (8).[31] Then he set about calculating – and now the sum starts to become a little bizarre – how many diameters of human capillaries fitted into the diameter of his own waist. He came out at some 19,602,000,000. Anyone wishing to check the calculation can take a look at the endnote.[32]

Because Antoni believed that the microscopic animals had similar proportions to humans, he concluded that approximately the same number of capillaries would theoretically fit into their minuscule 'waists'. And because – as he had described earlier – 300 of the little animals fitted end-to-end in the length of a grain of sand, the number of their capillaries that fitted into a grain was 300 × 19,602,000,000 = 5,880,600,000,000, almost 6 trillion.

That was a strange last step, as he took a geometrically irresponsible leap from a number that related to the capillaries in a two-dimensional plane (19,602,000,000) to one that related to length (300). And it could all have been much less complicated. If he simply wanted to determine the number of times the width of a capillary from one of his little animals fitted into the length of a grain of sand, he could have used a much easier sum: 33 (capillaries in a hair) × 600 (hairs in an inch) × 8 (inches in the diameter of his waist) = 158,400 human capillaries in the diameter of his waist – and thus just as many in the waist of a little animal. With 300 creatures in a grain of sand, that made 158,400 × 300 = 47,520,000 of their capillaries in the length of a grain of sand. Still an impressive number, but some 10,000 times less than the number that Antoni came up with.

But Antoni calculated even further with his immense 5,880,600,000,000. That number was 'so exceedingly great' that he needed a comparison to make it all clearer.[33] So he chose a number that everyone would consider immense: the number of times the thickness of a hair would fit into the circumference of the Earth. He determined that at 933,120,000,000.[34] A very large number, but according to Antoni, the number of little animal capillaries that fitted into a grain of sand was six times larger.[35]

Divining rods

Although it clearly didn't always work out, Antoni liked to use his research results to make calculations.[36] With his mathematical side, his fondness for counting, measuring and calculating, he was very much like present-day scientists, and especially as he combined that mathematical approach with accurate observations – but more of that later. Yet the occult still held a prominent place in the world Antoni lived in, and he was consequently confronted with problems that we would now consider unscientific, such as faith in the working of divining rods.

In 1696 Antoni received a letter from Pieter Rabus (1660–1702), a friend from Rotterdam who published the scholarly journal *Boekzaal van Europe*. Rabus told him of something curious that had happened to him. He had been visited by a man known as

K.V.B., probably Cornelis van Beughem, a lover of books and maps and publisher of *Stedenwysers*, tables showing distances between places.[37] He had told Rabus that he could use forked twigs – preferably from hazel – to find gold, silver and other minerals.[38] K.V.B.'s son, 'a youth of 18 years old', had inherited the talent from his father and had even given demonstrations: Rabus and a number of other guests had hidden gold and silver and the son had gone looking for them, holding the forked ends of the stick in his hands and the other end pointing out in front of him. While he was still at a distance from the hidden treasures nothing happened, but as he got closer, the stick began to 'move, turn, bend, and point with the tip either forwards or backwards' to the hiding place. The movement was so violent that the stick 'hurt his flesh, and almost wring off the rind'.[39]

Rabus considered himself a mortal enemy of suspicion and, in his time, divining rods were linked to demonic practices and 'witchcraft'.[40] Normally, he would have had serious doubts about such a story. But he had seen with his own eyes that it was possible. The son had done it at least 25 times, in the presence of many eyewitnesses.[41] What's more, Rabus's wife, Elisabeth Ostens, also proved proficient with the rod.[42]

And Rabus was not alone. A few years earlier, fellow enemy of superstition Balthasar Bekker had written that divining rods worked. Not only to find precious metals, but water and even murderers.[43] According to Bekker, there were natural explanations for the successes with divining rods, based on particle theories of thinkers like Descartes.[44] The idea appealed to Rabus, and he wrote in hope to Antoni that there may be something in the rods that caused a similar reaction to that of a magnet when in the vicinity of iron. Or perhaps the behaviour of the rod had something to do with particles in the bodies of K.V.B. and his son, and of others with a talent for divining. Rabus concluded that it could not be the rod itself, as then everyone could use one.[45]

Antoni's curiosity was aroused and he paid Rabus a visit, where he had had 'the great pleasure' of witnessing a divining demonstration.[46] He wanted to tell others what he had seen, but was afraid that they would not believe him. So he invited Rabus and his wife to his house, where Elisabeth demonstrated her talents to five other gentlemen, 'not once, but ten or twenty times'.[47]

For Antoni, Elisabeth's demonstrations were reason enough to start experimenting, and he found a fellow citizen of Delft who could also work with a divining rod. With this anonymous volunteer, Antoni conducted comparative tests with divining rods and magnets, to see if they worked in the same way. At first, he thought he had discovered a fundamental difference: while a divining rod moved towards precious metal (and not the metal towards the rod), it appeared to be the opposite in the case of a magnet: iron moved towards the magnet and not vice versa.[48] But after receiving a number of critical comments from the Royal Society, he changed his mind. A magnet could also move towards metal.[49]

Antoni then turned to the divining rod itself. He had the anonymous diviner experiment with twigs of willow, apple, pear and ash. They all worked but hazel was the best. So he compared the structure of the different kinds of wood under a microscope. He had conducted extensive research into wood before, and observed that it was full of veins. He saw now that hazelwood had an unusual amount of horizontal veins.[50] The implicit suggestion was that crucial particles could move freely through these numerous veins.

In the meantime, Rabus had also been working on a particle-based explanation and had a theory, which he published in his *Boekzaal*. It was a difficult explanation to follow: he claimed that the bodies of those with a talent for divining contained special particles, which were exuded outside the body. If they grasped a divining rod, the special particles moved through it. At the pointed end of the rod, they pushed coarse air particles away. At the same time, the precious metal also exuded particles, which in turn pushed coarse air particles to one side. If a divining rod came in the vicinity of gold, the space between was filled with special particles and few coarse air particles. And that, according to Rabus, made the divining rod move towards the metal.[51]

The explanation was dubious to say the least, but Rabus stuck by it. The whole divining issue had now become personal for him, as his wife was directly involved. Critics had loudly ridiculed his explanation. A group of men from Haarlem had accused the pair of deceit and had depicted Rabus in a pamphlet as a pompous, gullible and even fraudulent figure.[52] Deeply offended, Rabus published a poem in his journal attacking these 'libellous obscurants'.[53]

Miracle doctor

Antoni had by now also stopped experimenting with divining rods. For unknown reasons the anonymous diviner from Delft had lost his powers, and after that Antoni no longer seemed interested in the subject.[54] He did, however, continue to experiment with magnets and wrote to the Royal Society describing his progress.[55]

In the meantime, however, he was presented with a new mystery to occupy his mind, again via Rabus. This time, the protagonists were a 'miracle doctor' called Henricus Georgius Reddewitz (or Rettwich) de Rodachbrun and Rabus's publisher Pieter van der Slaart. The latter had developed an abscess on his leg following an accident.[56] A surgeon had tried to cure it by making an incision in the leg, but that had only caused pain and fever, and pus had to be drained from the wound every day.

The leg was in a bad way, but fortunately Reddewitz was there to save the day. He promised to heal it using a 'sympathetic secret'. Van der Slaart had to provide a sample of urine every day, which Reddewitz then treated with a powder without the patient even needing to be present.[57] The treatment appeared to work, and very quickly. Each day, Van der Slaart felt better and better and the pus disappeared. In no time, his leg felt as if the accident had never taken place.[58]

Van der Slaart had the miracle treatment recorded in a notarial act, a common way to document such matters in the seventeenth century. This was performed by Rabus who, besides producing his magazine, held a number of other positions, including that of notary. The act was published in the *Boekzaal*, followed by a statement by two Rotterdammers confirming the account.[59]

Strangely enough, Rabus himself was not overly enthusiastic about the texts, even though they appeared in his own magazine. 'What can I say?', he began in a postscript. 'This is not the first time that people have drawn up legal statements claiming that they have been cured of serious complaints after their urine had been treated in this way.'[60] And elsewhere in the same issue, he published a heated refutation by Rotterdam doctor Herman Lufneu of the curative power of the 'so-called sympathetic effect'.[61]

By the 'sympathetic effect', Lufneu meant a remote medical impact, where a patient was healed through the treatment of something outside their body – their urine, for example, or a sword that had wounded them. For the purposes of his argument, Lufneu supposed that the effect was possible, thanks to the kind of minuscule particles that Antoni was searching for and to which Rabus attributed the working of the divining rod. That would mean, Lufneu reasoned, that particles from, for example, the treated urine could be transported to the patient and thus have a curative effect. But that was difficult to imagine if there was a considerable distance between the patient and his urine, and especially if there were buildings in between.[62]

Could the principle perhaps be attributed to tiny animals such as those Antoni described around that time? Perhaps they lived in the urine and captured the powder that Reddewitz used. Then they would have to fly back to where they came from – the patient – carrying the curative powder with them. But this could only be possible, Lufneu reasoned, if the animals were able to orient themselves. And that would only be possible if they could smell the trail of the urine that had been brought from the patient to Reddewitz some hours earlier. But Lufneu again found it inconceivable that the trail would still be traceable hours later.[63] That led him to conclude that sympathetic medicine was nonsense. And Rabus agreed, as he made clear in a later issue of his journal. That he believed in the working of divining rods did not mean that he was prepared to defend the 'so-called sympathetic effect'.[64]

Antoni was aware of the whole discussion – either from Rabus directly or from the *Boekzaal* – and he shared the opinion of Rabus and Lufneu. He arranged a meeting with the miracle doctor, to inform him of his views on the matter 'as a Hollander, who is not accustomed to flatter'.[65] How was it possible, he asked Reddewitz, for the pus from Van der Slaart's leg to simply disappear? It couldn't just vanish into nothing. If it was gone, it must have been carried away in one way or another. And, according to Antoni, that could only have been done by the minuscule capillaries that he had seen in the human body. But, he said with his land surveyor's cap on, pus was far too thick to flow through the tiny veins.[66]

Reddewitz insisted that he was right and showed Antoni some small pieces of stone. He said they were the remnants of kidney or bladder stones, that he had pulverized using his sympathetic powder. Antoni didn't believe him, as he himself had once conducted an *in vitro* experiment with a kidney stone in very concentrated wine vinegar. He had left it for a year, after which the stone remained unaffected. That was evidence enough to convince him that the stones could not be destroyed with medicines, and certainly not at a distance.[67]

And how did Reddewitz think that sympathetic medicine might work? 'There are more things in nature which we cannot explain,' was his reply to that question.[68] To Antoni, who tried to understand the world by observing, counting and calculation, that was an unsatisfactory answer. He was willing to research subjects, like the divining rods, that reeked of superstition. But if an idea did not fit in with this particle-based theory, he lost interest – whether it was occult healing or Newton's theory of gravity.

He only wished to study what could be observed and measured. With one important exception: the small building blocks at the basis of everything. The more he looked, the more he became convinced that the smallest particles of all could not be observed, not even through a microscope. Take, for example, the capillaries in his little animals. He had never seen them with his own eyes, but he was certain that they existed. And those capillaries contained water particles that were even smaller, he thought – so incomprehensibly minuscule that he assumed that no one would ever be able to see them, no matter how far science progressed. But that they existed he was never in doubt.[69]

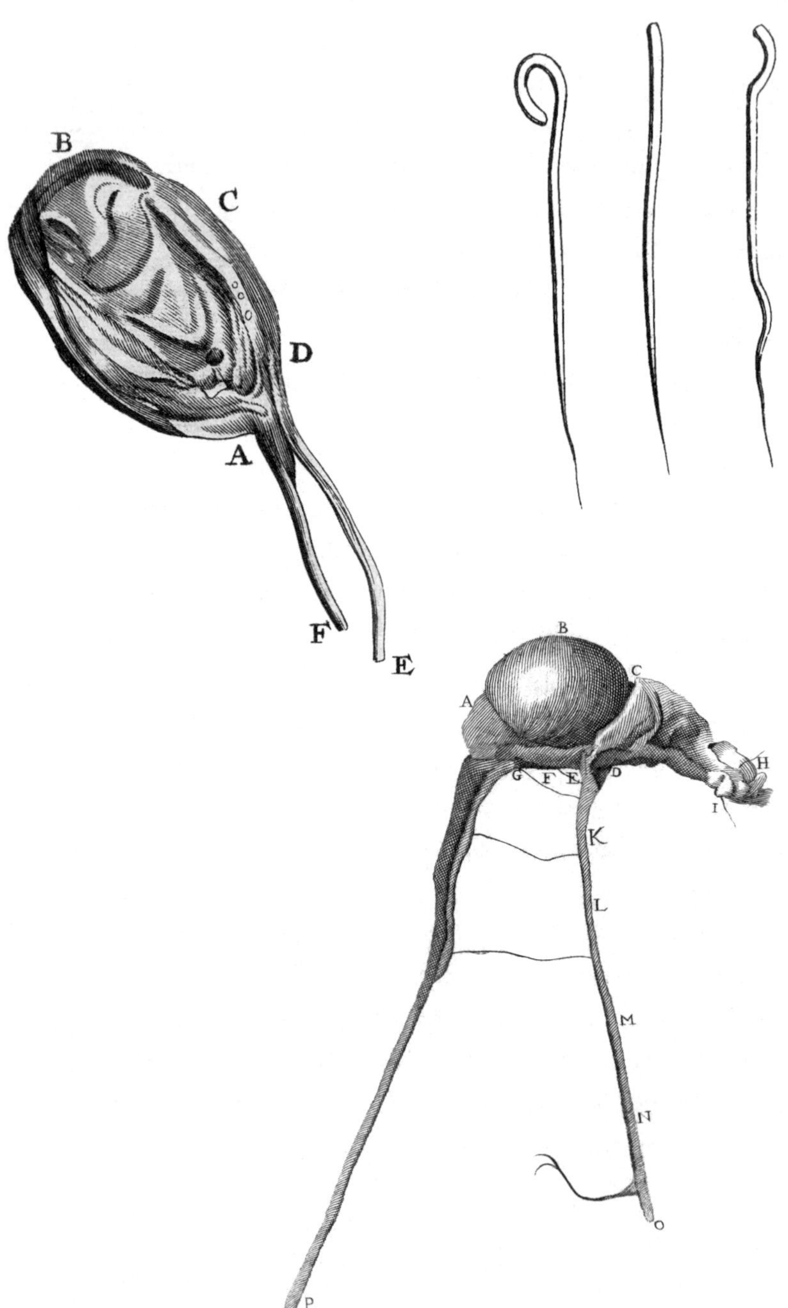
B
C
D
A
F
E
B
C
A
H
G
F
E
D
I
K
L
M
N
O
P

FIVE

Little Animals and the Most Addle-Pated Propositions among Physicians

It was the breeding season for rabbits and Antoni wanted to witness the 'union' between a buck and a doe. A breeder he knew tipped him off that the rabbits were 'in rut', so he went to watch in the hope of collecting a sample of the semen that the buck discharged.[1] During the first mating he attended, a few drops of thin fluid escaped from the doe, but he was unable to capture them in time. Fortunately, the rabbits mated again a few minutes later and this time he successfully gathered a sample of the fluid released by the doe. He took it home and examined it through a microscope, but the results were disappointing.[2]

So he visited the breeder again, and this time had a buck and a doe placed together in a wooden cask to make it easier to gather any fluid that may escape. This time, however, the doe 'was not at all inclined to copulate' and kept her 'hind part entirely retracted'. The buck, however, refused to give up and was, in Antoni's words, 'very hot', making five or six attempts to mate, resulting in the hind parts of the doe becoming 'very wet'.[3] Eventually, the buck released his seed, which was 'very thickish'. Antoni collected it immediately and, 20 minutes later, examined it through a microscope.[4]

*

When Antoni imposed himself on the mating rabbits, in the late winter and early spring of 1678, he already had an idea of what he might see when he observed the semen through a microscope. He had examined semen before, from humans and dogs. The first time was in 1674, when Henry Oldenburg, secretary of the Royal Society, asked him to study the semen of a man.[5] Then he had only seen globules and because, in his words, 'he felt adverse from making further investigations', he had discontinued his observations.[6]

In 1677, his interest in semen was reawakened when he received a visit from a student, Johan Ham.[7] Ham had used a microscope to examine the semen of a man who had contracted gonorrhoea after having sex with 'an unhealthy female individual'.[8] The semen contained active little animals with tails.[9] Ham had discussed this unexpected result with a member of his family, Leiden professor and follower of Descartes, Theodorus Craanen. Craanen had advised him to consult Antoni, and the two men consequently examined a sample of the patient's semen in a small glass phial and saw the little creatures swimming around.[10]

Antoni wanted to know whether the creatures were related to the gonorrhoea, or whether they were also to be found in the semen of healthy men. He determined that by using his own, healthy semen.[11] He was at pains to make it clear that he had acquired the semen in a natural way, 'without sinfully defiling' himself, collecting only what remained 'as a residue after conjugal coitus'.[12] He performed that conjugal coitus with Cornelia Swalmius, whom he married on 25 January 1671, more than four years after the death of his first wife, Barbara.[13] Cornelia was a spinster of 36 years.[14] She and Antoni had no children, so Antoni's only surviving child was his daughter, Maria, who was fourteen when Cornelia became her stepmother.

Her husband's research into semen was probably not very enjoyable for Cornelia, as Antoni wrote that he gathered a sample and was observing it within six pulse beats of ejaculation.[15] Just as in the semen of the gonorrhoea patient, he saw animals smaller than the red globules in blood. He estimated that more than a million would fit into his standard unit of measurement, a grain of sand.[16] Their tails were five or six times as long as their bodies, with a thickness of about 1/25th of the body. They used the tails to move forward, like a snake or an eel swimming in water.[17]

Antoni considered the creatures to be 'marvels of nature'.[18] But it was not the first time he had observed life forms of such small dimensions. That had occurred before, in the summer of 1674,

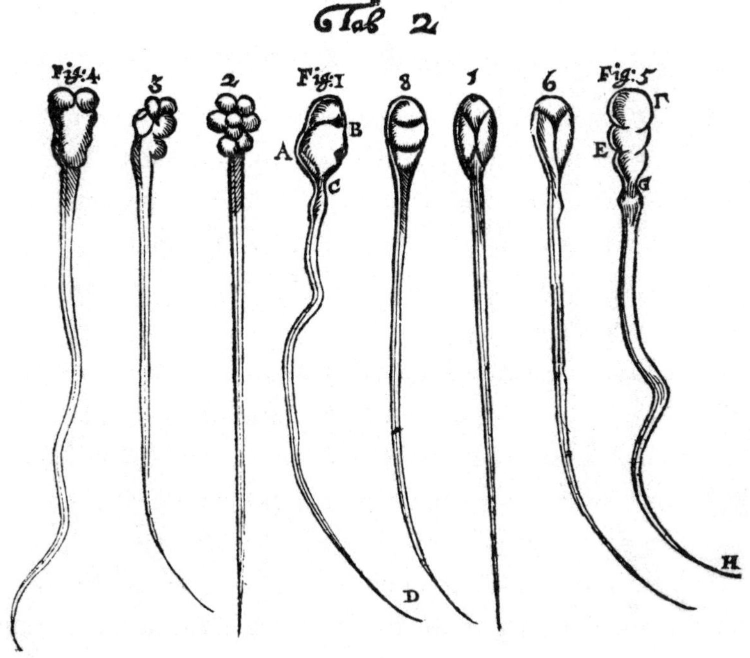

23 From a drawing by Antoni of the semen cells of a human (figure 1 alive, figures 2 to 4 dead) and of a dog (figure 5 alive, figures 6 to 8 dead).

when he had visited the Berkelse Meer. With hindsight, he had made an imposing discovery, but anyone reading his letter on the matter could easily miss it. First he wrote pages about the structures in the eyes, nerves, kitchen salt, his early observations of English chalk and minute grains in clay.[19] Only then did he, in what for him were concise terms, describe the life that he saw in the water he had taken from the lake: the little particles joined together by spiral-shaped green strands as thick as a hair, and round, oval and elongated minuscule animals. He wrote that the oval creatures had legs near their heads and fins on their backs. Some had scales, while others had none. And most of them moved swiftly in all directions, except the elongated ones, which were lazier than the others.[20]

Parrots' beaks and tortoises

Generally speaking, Antoni would show it if he was enthusiastic about his observations. He would write with wonder about the delightful, exceptional or incomprehensible things he had seen.[21] But in his letter about the Berkelse Meer, his tone was quite reserved.[22] He only returned to the subject a year later, and that was a matter of coincidence.

He was actually studying air, probably with the kind of experiments described in Chapter Three, with a glass tube and a plunger.[23] He was using water in one of the experiments, and when he looked at it through a lens, he saw the 'most wretched' creatures he had ever seen. They were flat on one side, except when they moved. Then two 'little horns' protruded, which reminded Antoni of moving horse's ears. The rest of the body was somewhat round, but tapered towards the back, ending in a tail as thin as a spider's thread, with a globule at its tip.[24] When the creatures moved their tails became entangled, so that they had to make great efforts to free themselves again. Antoni found that pitiful to see.[25]

Besides these lamentable horned creatures, there was more life moving around in the water, in all shapes and sizes. This time, Antoni was so fascinated that he set up a series of observations. He sought the microscopic beings, in all possible variations, in raindrops, water from the well in his garden, the canals in Delft and the North Sea.[26]

They were incomprehensibly small: in rain, and in water that he had used to soak peppercorns to study the sharp taste of pepper, he saw little animals, 10,000 of which would fit into a grain of sand.[27] There was so much diversity that Antoni grew tired of trying to classify them all. When he found new types of creatures in canal water, he decided that describing how they all looked and moved would be 'too tedious'.[28]

He usually did his best to give his readers a good impression of all the minuscule creatures and objects that he observed, often drawing comparisons with things from the big world, such as above with horse's ears. In well water with pepper, for example, he saw creatures that were partly oval-shaped, but which had a curved bend at the front, 'like a parrot's beak'.[29] In water in which ginger had soaked for some time, he observed minuscule creatures that hopped like magpies.[30] Yet other creatures were round on top and flat underneath, like tortoises.[31]

Antoni's portrayals were creative, but modern readers may wish to have a little more certainty. That was in any case true of Clifford Dobell, a microbiologist who described Antoni's work in 1932 from his own scientific perspective and wanted find out exactly which micro-organisms his predecessor had seen.[32] He tried to link Antoni's seventeenth-century classifications to twentieth-century

24 In water, Antoni saw roundish creatures with flat upper sides, that tapered at the back to thin tails. They were probably *Vorticella* or 'bell animalcules' (as shown here).

microbiological terms. The creatures with horns like horse's ears, for example, were clearly single-celled *Vorticella* or bell animalcules (see illus. 24).[33] That would mean that the 'horns' were illusions created by the movement of the cilia, the vibrating hairs around the top edge.[34]

Dobell also tried to find a scientific explanation for the very first micro-organisms that Antoni had seen, the green particles connected together by trailing strands in the water of the Berkelse

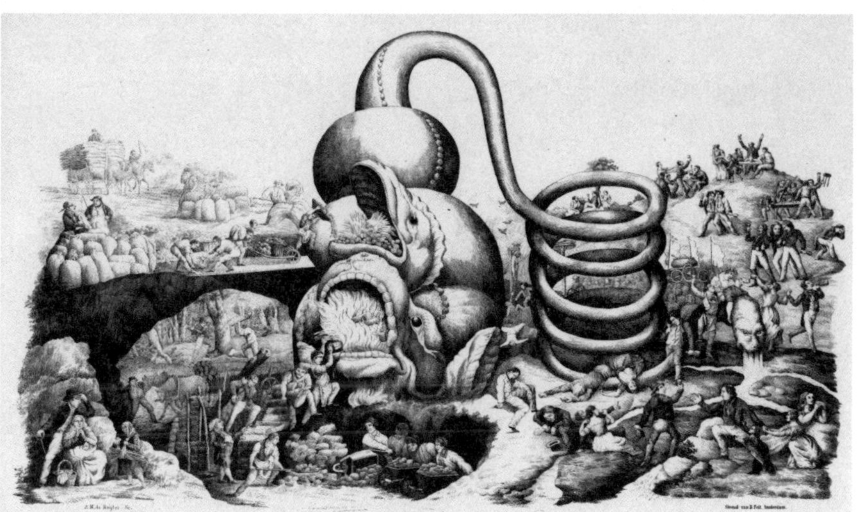

25 In the water from Berkelse Meer, Antoni saw green particles with connected strands that reminded him of the spirals that distillers used to cool their fluids, as seen here on an anti-alcohol print from the 19th century by Adrianus Martinus de Ruyter.

Meer. The strands, Antoni had written, were twisted like the spirals that distillers used to cool their fluids. Dobell concluded that these were *Spirogyras*, organisms from the genus of green algae. But, almost a century later, micro-photographer Wim van Egmond disagreed. The spirals in *Spirogyra* are located within the strands, while Antoni wrote that the strands themselves were curled into spirals. Van Egmond considered *Dolichospermum* a more suitable candidate. That is a kind of blue algae, which join together and curl up, and are mainly visible in the middle of summer. They then form whitish clumps, and that is precisely what Antoni saw when he visited the lake during the summer.[35] That would mean that Antoni had not only seen curious little creatures, but bacteria. Despite the somewhat confusing name 'blue-green algae', *Dolichospermum* falls into the category of bacteria, cyanobacteria to be precise.

26 According to Clifford Dobell, the first micro-organisms that Antoni observed were of the genus *Spirogyra*.

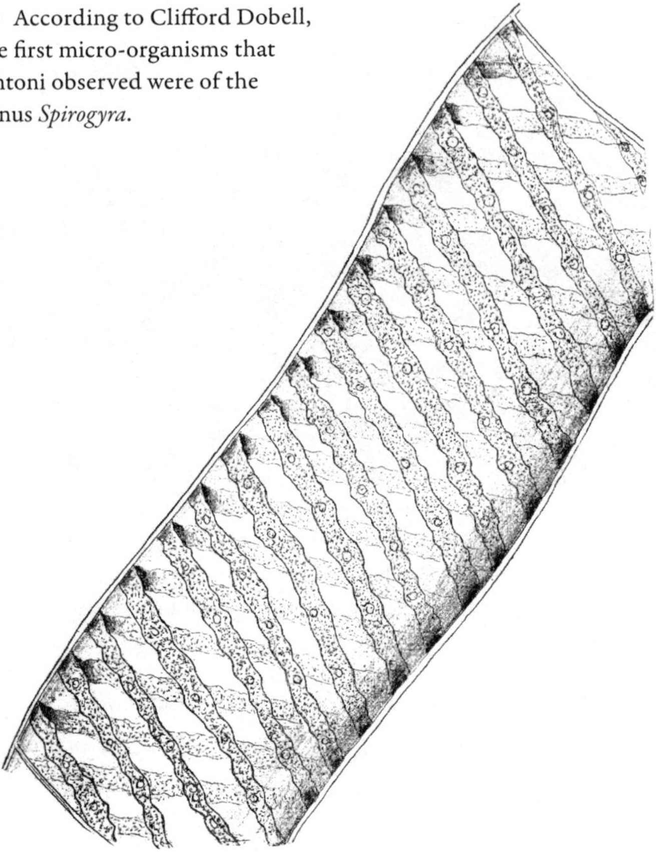

27 In the Berkelse Meer, Antoni saw green, spiral-shaped strands. It is most likely that they were blue-green algae of the genus *Dolichospermum*, like the *Dolichospermum circinale* pictured here.

All this is fascinating to know from a natural scientific perspective, especially for microbiologists who see Antoni as a founding father of their discipline. But anyone wishing to understand Antoni's world better would be well advised to steer clear of these kinds of considerations. He knew nothing of the names that those who came after him would give to his little animals and the categories they would be placed in. And, more importantly, he had no notion of the composition of the creatures, how they lived, and the impact they had on their environments. The concept of 'single-celled' would have meant nothing to him. He saw the little animals as amusing smaller variants of the living beings he could observe with his naked eye. They were interesting to look at, but that was all.

With the 'little seed animals', it was a different story, even though it took a while for Antoni to give them his full attention. As was by now his custom, he announced the discovery of the little animals in a letter to the Royal Society. He had the letter first translated into Latin, the language of scholars, while he usually sent his descriptions to London in Dutch, where someone would translate them into English. He apparently wanted to make his sex-related research sound a little more authoritative. He also added some words of caution: if Royal Society president William Brouncker, to whom he addressed the letter, thought his observations might disgust or scandalize the society's members, Antoni requested him to destroy it.[36]

Brouncker did not destroy the letter, but forwarded it to another Royal Society scholar: doctor, plant anatomist and microscopist Nehemiah Grew (1641–1712). Grew found Antoni's observations fascinating, but also had his doubts about their accuracy and asked for more research from other animals than humans.[37]

Antoni honoured Grew's request, observing the fluids from mating dogs and visiting the rabbit breeder. Interestingly enough, both animals proved to have similar little seed animals. That was pleasing, but Antoni also hoped to find something else that he had seen previously in his own semen: a tangle of small and larger vessels that, according to him, comprised a large part of the semen. In line with his particle theory, he assumed that the semen contained building blocks for a future baby. Because of their shape, he suspected that the vessels were the forerunners of nerves and blood vessels: all neural pathways and blood vessels in the human body were thus much enlarged versions of the vessels in the semen of the father.[38]

Grew doubted the existence of the vessels, but Antoni was convinced that his observations were correct, and that he would also see them in animal semen.[39] Sometimes he caught a glimpse of them, but often his quest was difficult.[40] He did not see them in the semen of dogs and it took two attempts to discern a few in the case of rabbits. And they were very unclear. He wrote that, if he had not seen them before in his own semen, he would have missed them.[41]

In hindsight, that is logical, as such vessels do not belong in semen, in humans or other animals. That begs the question why Antoni thought that he saw them, in the first instance in his own semen. Were they perhaps an optical illusion, caused by his way of working? That is possible, but when the researchers from Visualizing the Unknown observed human semen under similar conditions, they saw the little animals but not the vessels.[42] So why Antoni saw the vessels remains a mystery.

Antoni himself approached the problem from the opposite direction, and tried to explain why he no longer or hardly ever saw the vessels in his follow-up observations, while he was convinced that they must exist. He blamed the conditions under which he studied his specimens. He collected his own semen at home and could examine it under a lens immediately. But the animals mated elsewhere and he did not have his instruments to hand. In the time it took to examine the semen, its original components had become unrecognizable.[43]

In the following months, despite the disappointing observations, Antoni attributed more functions to the vessels. While he at first assumed only that they contained smaller versions of nerves and blood vessels, six months later he suspected that they contained

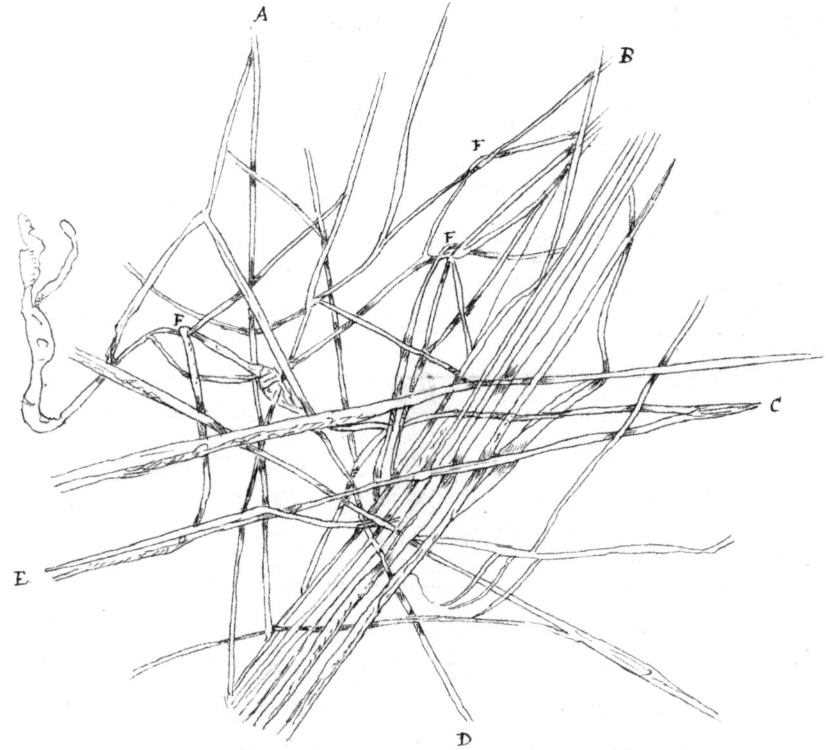

28 Besides the little seed animals in semen, Antoni also saw the vessels shown in this figure, which he believed contained the components for future babies.

all building blocks of future babies. He even believed that the little seed animals, whose role was not yet clear to him, were initially to be found in the vessels. During mating – and certainly in the case of rough and difficult attempts such as those of the male rabbit in the wooden cask – the vessels could be ripped to shreds, releasing the little animals. The remains of the vessels were then no longer to be seen.[44]

Dogs, tomcats and cockerels

He was of the same mind when he decided to approach the subject from a different perspective. Instead of waiting for the animals to mate in his presence so that he could collect their semen, Antoni studied the testicles of various male animals, including dogs. To do that, he sought the help of a labourer who earned a living castrating dogs and cats. The man brought two dogs to Antoni's house and removed the testicles from their – living – bodies, so that Antoni could examine them, and see that the sperm ducts were full of little animals.[45]

That was a positive result, but the observations were hampered by the fact that blood was released when the testicles were cut. So Antoni adopted a different strategy in a subsequent experiment, one that was even more distressing for the test animal. This time, Antoni chose a cockerel that had been kept from mating for a few days, in the hope that the testicles would be full of semen. First he cut through the neck to release a large quantity of blood, to give him a clear field of vision. Then, while the cock was still alive, he made an incision in the hind part of its body and removed the testicles, together with the sperm ducts. In the ducts, he saw 'such an enormous number of living animalcules that I was quite amazed'.[46]

It was clear, he concluded, that the testicles were intended to form the little seed animals and to keep them until they were 'ejected'.[47]

That still did not answer the question about the vessels he had observed previously. He had not seen them in the testicles, but that did nothing to shake his belief that they existed and that they contained the little animals.[48] He needed to conduct further research, but no longer with dogs from the castrating labourer. Antoni had heard that the man had acquired his two test animals in a dishonest way. One had fleas, and the man had promised the owners that he would get rid of them, without saying anything about castration. The other he had simply taken off the street, without seeking the consent of a possible owner. Antoni did not agree with 'such doings'.[49]

*

Antoni conducted further experiments with different kinds of animals, always with the same result: a lot of little animals, but no vessels to be seen. He consequently said nothing more of the vessels in his letters, focusing attention on the little seed animals. And in early 1683, more than five years after Ham's visit, Antoni wrote that he had been mistaken. There were no vessels in semen. It was the little animals that were important.[50]

Sometimes, he considered the idea that little seed animals contained preformed miniature humans or animals.[51] But if these existed, they were too small to be observed. Moreover, the bodies of the little seed animals did not look like the creatures that would grow out of them, leading Antoni to propose a slightly different hypothesis: the little seed animals contained all the bodily parts that a human or animal would possess at birth, but in minuscule form.[52] The various components were not yet in their correct places in the seed animals, but were rearranged early in pregnancy, forming a recognizable creature that then only had to grow into a viable young animal.[53]

Female 'testicles'

Female mammals thus supplied no body parts for their young. According to Antoni, they provided only nutrition for the semen and a place for it to grow.[54] That was a remarkable statement, especially given that, a few years earlier, medical researchers had attributed female 'eggs' a central role in reproduction.

We have already come across one of those researchers: Regnier de Graaf, who saw in the female 'testicles' (ovaries) the vesicles that we now call 'Graafian follicles', and which contain egg cells (ova). But he was not the only one to study the vesicles. Various physicians took an interest in them in the 1660s, including a former professor and a fellow student of De Graaf, Jan van Horne and Jan Swammerdam.[55] They had discovered long before De Graaf that the vesicles were or contained eggs, but had not succeeded for many years in publishing their research.[56] And so it happened that, in 1671, both De Graaf and Swammerdam – Van Horne had died in the meantime – were on the point of publishing their findings. That led to a heated dispute between them and insinuations from both sides that the other had stolen their ideas.[57]

Despite all the mutual bad feeling, De Graaf and Swammerdam agreed on one important point: the female 'balls' and the follicles were essential for reproduction. Antoni, however, thought differently, and explained why in letters to the Royal Society.

First, he disputed a statement made first by William Harvey (who had described the blood circulation system) and then by De Graaf. They had cut open the reproductive organs of female animals after mating and examined them. According to them, there was no semen (that is to say, no white fluid) to be seen.[58] Clearly, they reasoned, the semen did not penetrate into the female sexual organs. Or, if it did, it was only in the form of super-fine, invisible particles.

But Antoni was not overly concerned by the fact that Harvey and De Graaf saw no evidence of semen. They had tied the female animals down and cut into their sensitive parts. That had undoubtedly caused them great 'fright', and everyone knew that 'anxiety, fright and uneasiness' can cause abortion in a pregnant female. During these horrific experiments, the female's body would most certainly try to expel the seed, together with the excrement and urine that animals void at such moments – as Antoni knew from experience. In addition, under such circumstances, the body would be unable to provide the nutrition that the little seed animals needed to survive. Furthermore, Antoni had learned that semen changed into a watery matter when exposed to air, which was inevitable when the womb was cut open. It was therefore not surprising that Harvey and De Graaf saw nothing that they recognized as semen.[59] But if they claimed that the eggs were paramount and that the semen does not enter the womb, then they were propagating 'one of the most addle-pated propositions among physicians'.[60]

To demonstrate this, he had a female dog killed after she had mated three times. He clearly wanted to avoid her experiencing anxiety, fright and uneasiness and expelling the seed that she had received during mating, so he had her put to death with a dagger into the spine.[61] He then inspected her 'tubas' (that is, her fallopian tubes). At first, he saw nothing with his naked eye. But when he looked through a magnifying glass, he discovered to his satisfaction 'an abundance of living animalcules'. These were little seed animals, and they proved to be located in the womb.[62] On this point, then, De Graaf and others who shared his opinion were wrong.

Watery eggs

Antoni had also examined the alleged eggs of cows and lambs, but he was not impressed. In the first place, he saw enormous differences between the vesicles. Some were very small, others larger, and one was the same size as the other 'testicle' (that is, ball or ovary). And they were of different colours – white, yellow and red.[63] He did not explain why he considered that a problem, but he apparently thought it illogical that vesicles with so many differences could have the same function. Furthermore, he concluded from a later, more detailed inspection that 'eggs' had widely varying shapes and did not look like eggs at all.[64]

The eggs also contained a watery substance, mixed with hardly discernible globules.[65] There were no building blocks for a baby to be seen. And to make it even less feasible that the eggs played a role of any significance, they were enveloped in two membranes that were stronger than the vesicles themselves.[66] De Graaf had written that the eggs came out of the balls (what was later to be called ovulation) leaving the ball behind empty. But Antoni found that unlikely, given the strength of the membranes. He concluded that De Graaf had seen a 'defect of nature'.[67] That was confirmed in his eyes a few years later, when he tried to remove an 'egg' from a vesicle with his fingernails. The vesicle was so firmly attached that he tore the whole ovary to pieces.[68] It was therefore impossible, Antoni wrote, for the tuba to suck or pull the egg from the ovary.[69] No one had claimed that this happened, but Antoni announced that many 'learned men' had made such allegations, while he thought it through logically.[70]

And then there was the argument of Antoni the land surveyor. During his research into female 'balls', he had seen that the 'eggs' could be quite large: a big as a pea or, in one extreme case, as a vesicle itself. At the same time, the channel in the tuba was as thin as a needle. That made it implausible for the 'eggs' to pass through the tuba to the womb.[71] They would simply not fit.

This was a somewhat dubious conclusion, given that De Graaf had already shown that the vesicles served as an outer shell for the eggs, which were themselves much smaller. Moreover, other researchers had seen eggs in the tuba. But Antoni dismissed that as nonsense. He proposed that they were little seed animals or other residual particles.[72] He thus had an answer to every objection to his own opinions.

*

For Antoni, thus, the focus had to lie on the little seed animals. After coitus, they would seek out a spot in the womb where a blood vessel was close to the surface, so that they could feed off it. Antoni thought these spots must be rare, otherwise a mother would give birth to many more babies than was the case. Once a seed animal had settled itself in a good spot, it would lose its tail and take on the oval shape of the curled-up embryo in the womb.[73]

Antoni now confused his argument a little more by writing that, around the time that the little animal had found itself a suitable spot in the womb, it also acquired a soul. A sentence later, he suggests that the acquisition of a soul is related to the movement of the little animals.[74] This fits in completely with his conviction that the lively little seed animals were of great significance to reproduction and the immobile eggs were not.[75] But why did he assert that the little seed animals acquired a soul at the moment that they were settled in the womb and were thus immobile? That is because he would otherwise have to say that all of the moving seed animals had souls, implying that thousands of souls would die with every ejaculation. It was difficult to imagine that God in all his goodness would allow that to happen, but it also made Antoni's explanation more than dubious.

In addition, Antoni had still not explained what function the vesicles served if they did not provide any building blocks for the babies. Nourishment was one option, but he rejected that idea with the argument that an embryo could never grow larger than the vesicle that fed it. And that was clearly the case.[76] And thus, he wrote in March 1694 – more than sixteen years after Johan Ham's visit – that the female balls and eggs had no function. He found that an unsatisfactory conclusion, as he felt that nature was providential and made nothing without a reason. But there were exceptions to the rule. Nipples, for example, had a very clear function in the case of women, while male nipples were completely superfluous. The reverse applied to male and female balls: the former were essential to reproduction, while the latter had no use at all.[77]

Enemies

Antoni would adhere to this standpoint until his death, but he received little support.[78] Only one fellow microscope enthusiast agreed with him. And he was furious with Antoni, because he wanted the honour of discovering the little seed animals himself.

Nicolaas Hartsoeker (1656–1725) became interested in microscopy at an early age. In his teens, his father took him to see Antoni, who frequently received visitors curious about his work.[79] Encouraged by his father, Hartsoeker had examined a number of Antoni's specimens.[80] Some years later, a mathematics teacher taught him to make lenses by holding a glass thread in a flame until it curled up into a ball. Hartsoeker placed the ball between two metal plates, so that he had a microscope similar to those used by Antoni.[81]

In the spring and summer of 1674, as he himself wrote, Hartsoeker observed everything he could lay his hands on, including human semen. He claimed that he saw – three years before Johan Ham – that the semen was full of little animals. But he thought that they were indications of a disease and, what is more, had no one to discuss them with. He had therefore done nothing more with his findings.[82] He later went to study in Leiden, where he was taught by Theodorus Craanen, who regularly visited Antoni in those years and who sent Johan Ham to Delft in 1677 to talk to him about the little seed animals.[83] Craanen was thus interested in microscopy, but Hartsoeker himself did not get the opportunity to work with lenses during his time in Leiden. He only returned to that – again according to his own account – in early 1677, after he had given up his studies.[84] He had once again examined his own semen and seen little animals, which reminded him of tadpoles. This time, he had shared his findings with his old mathematics teacher and with a friend. Together they had extended the research to other animals – a dog, a cockerel and a pigeon – and each time, they saw the little animals. They did not all look the same: the animals in the bird semen did not have heads like tadpoles. They looked more like worms, or eels. But no matter what they looked like, Hartsoeker was certain that they were a natural component of semen. And he did all this, according to his own account, months before Antoni had seen them.[85]

There is reason enough to doubt the verity of Hartsoeker's account, as he did not always tell the truth. Years later, he wrote an essay unjustly accusing Antoni of stealing other people's ideas.[86]

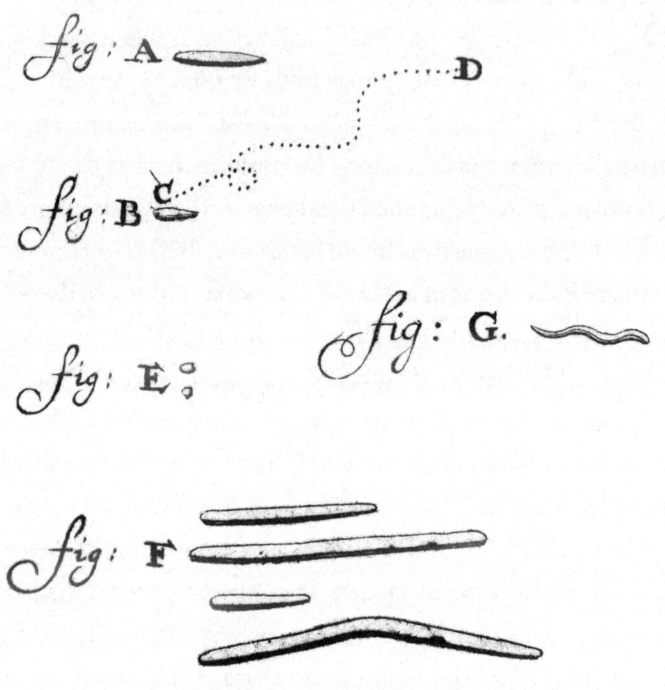

29 Antoni observed moving particles in a mixture of tartar, saliva and clean water. He showed with a dotted line that they moved 'very prettily'.

Hartsoeker claimed that, around 1677, he had discussed the little seed animals with a fellow Rotterdammer. They had used the word 'saliva', which was, according to Hartsoeker, a euphemism for semen in Rotterdam dialect. That had led to an erroneous story doing the rounds that Hartsoeker had seen little animals in saliva rather than in semen. Antoni had then, not wishing to be outdone, claimed that he had seen countless little animals in saliva, while anyone with a lens could see – according to Hartsoeker – that that was not true.[87]

Hartsoeker was far wide of the mark with these allegations. At the time they were studying semen, Antoni was convinced that saliva contained no living beings, only 'globules'.[88] He had written about that before and, after discovering the little seed animals, he had confirmed it.[89] He had thus not let himself be led astray by the Rotterdam saliva story. He did not change his mind until a year later, in 1683, when he took another, closer look at the contents of his own mouth. Antoni was proud of his teeth, which he rubbed clean with salt every morning. After eating, he would remove any pieces of food with a toothpick and rub them with a cloth.[90] That ensured that his teeth looked clean and healthy, especially for a man of fifty. But when he observed them closely through a magnifying glass, he saw a little white matter in the gaps.[91] When he examined the matter under a microscope, he still saw no living animals. It was only when he mixed the white matter with a little saliva and animal-free rainwater that he observed a large number of living beings.[92]

Hartsoeker's accusations about saliva were therefore not true. So what does that say about his allegation that he had seen little seed animals before Ham and Antoni? The first proof that he studied semen dates from 14 March 1678, when Hartsoeker wrote a letter to Christiaan Huygens about lenses and little seed animals. It was clear from the letter that they had spoken about the subject before.[93] That was a few months after Ham's visit to Antoni, but before the discovery of the little seed animals had been published. It is therefore possible that Hartsoeker made the same discovery independently, perhaps even early in 1677, as he had claimed. Or perhaps he was simply lying and had talked to Huygens about the little seed animals after hearing about Antoni's research.

It is no longer possible to find out exactly how and when everything happened – not even from the writings of Huygens, who was very closely involved in the developments. He gave only a number of vague hints. In the learned journal *Journal des sçavans*, he wrote an article about a new method of making lenses in Holland. That was the method that Hartsoeker had learned from his mathematics teacher and told Huygens about. Huygens also mentioned the little seed animals that had been observed with microscopes.[94] Hartsoeker took offence at this, feeling that Huygens should have mentioned him by name.[95] That led to an addendum in the same journal, stating that Hartsoeker had made a significant contribution to the new lens-making method, and that he had seen that the semen of cockerels looked different to that of mammals, having no rounded, tadpole-like head.[96] Huygens made no mention of Hartsoeker having taken the first great step in the research into semen, despite the addendum being printed at the latter's insistence.

It is tempting to assume that, in Huygens's eyes, Hartsoeker was not the first and that this honour belonged to Ham and Antoni. But Huygens was equally vague about them. He wrote to the Royal Society about the little animals in semen, stating that 'it was said' that Ham had discovered them.[97]

There thus remains a significant margin of doubt. But at the start of 1679, the Royal Society published Antoni's letter about Ham and the little seed animals. That was the first unambiguous publication and, consequently, Antoni and Ham were attributed with the honour of being the first as, at that time, publication weighed heavily in determining who had made a discovery. It was a defeat for Hartsoeker, who turned up at Antoni's house in a fury to demand justice. Antoni did not, however, wish to speak to him and refused to let him in. From that moment, Hartsoeker was to write later, the two became enemies.[98]

The issue flared up again in 1698, when a book by Hartsoeker was published. In the book, he claimed that he had started researching semen twenty years earlier, to his knowledge as the first to do so.[99] Antoni read the account and believed that Hartsoeker had made a mistake. If he had started examining semen twenty years earlier, in 1678, then he was clearly not the first, as Antoni had told the Royal Society about his own research in November 1677. He had also examined semen a few years earlier, at the request of the Royal Society, though then he had only seen globules.[100] Hartsoeker's claim thus appeared very strange to him.[101]

At the core of the dispute was the fact that the 1698 book was a translation of a French text by Hartsoeker published in 1694. The text also stated that Hartsoeker had started examining semen twenty years earlier.[102] That referred back to 1674, in the period before he started studying, when he thought that the little animals were indications of a disease. The translator had not changed the sentence in line with the later publication date, leading Antoni to dismiss Hartsoeker's claim, as it now suggested that he had started his research around 1678.

The two would never be reconciled. Antoni would always insist that he and Ham were the first to research semen, and that the little seed animals were infinitely more important than the eggs.

Either way, Antoni and Ham were widely honoured as the first to uncover the secrets of semen, and that was clearly a permanent thorn in the side of Hartsoeker. Many years later, he published another article in which he tried to convince his readers that he really had been the first.[103]

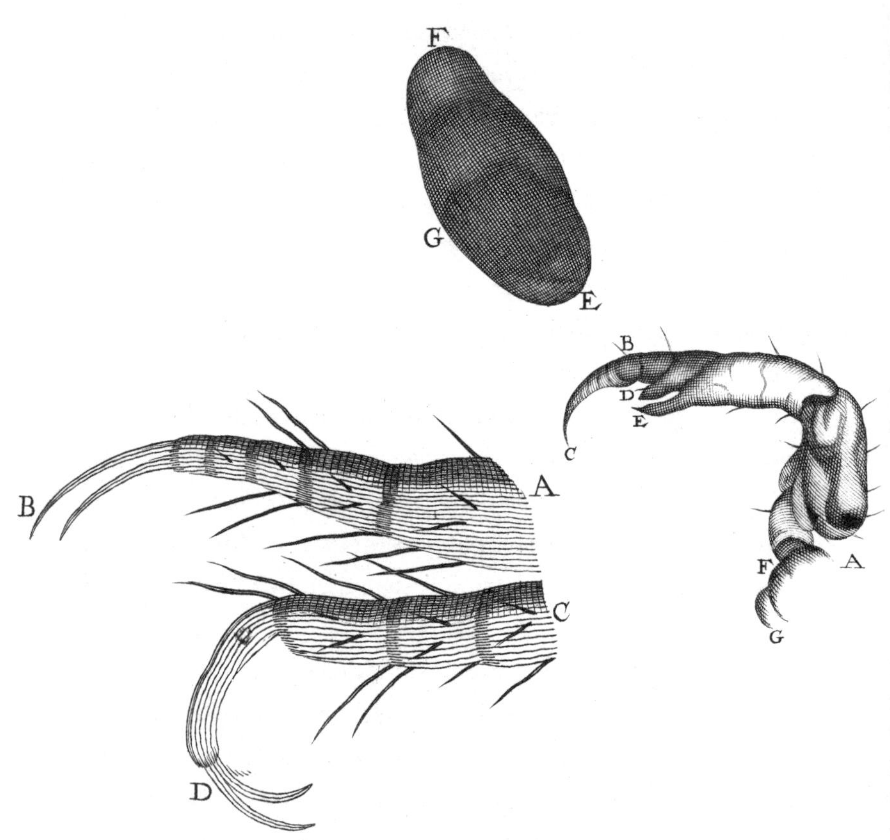
F
G
E
B
D
E
C
B
A
A
F
G
C
D

SIX

Openness and Smokescreens

Back to the present for a moment. A bee has flown into my house and is now lying dead on the window sill. It is a perfect opportunity for me to replicate one of Antoni's studies, which he sent to the Royal Society three-and-a-half centuries ago.[1] So I take a replica of one of his microscopes, one with a pin in front of the lens, and try to prick the pin through one of the bee's legs. But it's easier said than done. The fragile little limb sticks to my hands, gets lost under the desk. Bits fall off, but it steadfastly refuses to be pricked. Poster tack, that sticky stuff like chewing gum, helps somewhat, but I still have to get the leg the right distance from the lens. That's essential: a little too close or too far away and it is vague and unrecognizable.

Eventually, after more messing around, I finally get a sharp image. But it's too dark. So – as Antoni wrote that it helped – I put a candle next to my face so that the leg is illuminated from the side.[2] I see a small stalk, surprisingly shiny and covered in little twigs. The section of leg that the lens shows in sharp focus is tiny, so it's difficult to work out exactly what I'm looking at. While I'm trying to puzzle that out, I smell something scorching. It's my hair, hanging in the candle.

The damage is limited and I want to push on. Antoni didn't examine the legs of the bee, but its sting and protuberances near

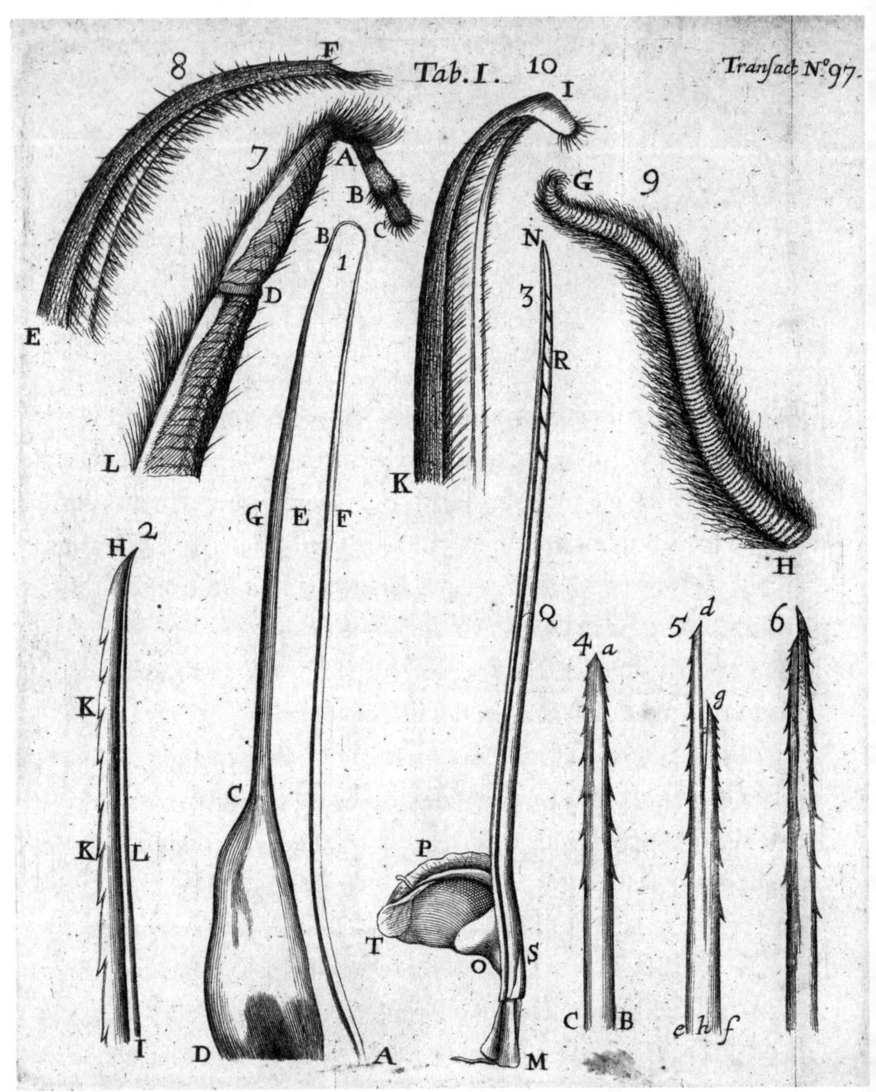

30 Antoni examined the sting (figures 1 to 6) and parts of the mouth (figures 7 to 10) of a bee.

the mouth. And they are a lot more difficult to detach. It's certainly beyond me. Perhaps I left the bee on the window sill too long before examining it. Or maybe I'm just too clumsy. But while I'm fiddling around and the bee falls apart, my admiration for what Antoni achieved grows even more. How did he do that?

*

Antoni conducted his microscopic work in the front room on the first floor of his house, Het Gouden Hoofd. It was originally a corner of his bedroom, partitioned off from the rest with a wooden wall. This created what was known as a 'comptoir', a small office where merchants kept track of their transactions.[3] It is quite possible that the small room dated from the time when Antoni still dealt in textiles and needed a quiet place to do his bookkeeping.

On days that he had no duties to perform at the town hall, he would often spend hours looking at little animals and other curious tiny objects.[4] Or he would work on designing new microscopes. He extracted the metal from ore himself or obtained it from a coppersmith.[5] He had no training in working the metal, but he had watched a silversmith and other craftsmen very closely. That had taught him to give the metal the desired hardness, drill holes and solder pieces together.[6] He also had a lathe in his comptoir, which he could use to make rods and threads, and to grind lenses.[7]

As his critical visitor Zacharias Conrad von Uffenbach noted, the small microscopes were not very refined.[8] But they worked well, and Antoni did not have the time to improve them endlessly, as he needed masses of them. Once he had succeeded in getting a sharp image of a specimen – a bee sting, a mosquito's wing, a piece of spider's web – through a lens, he would store it, microscope and all, in a lacquered box or case in a cabinet in the comptoir.[9] In this way, he accumulated the more than five hundred little wonders that were sold after Maria's death.[10]

He kept all his specimens with their own small microscope because the softness of the metal meant that there was a limit to the number of times they could be adjusted. But it also says something of how difficult it was for Antoni to obtain sharp images of his specimens through the lens. If that had been easy, he could have managed with a few microscopes with lenses of different strengths, making up a new specimen each time. That would have saved a lot of preparation work. But it was apparently not that simple – and therefore not so surprising that I failed with my bee.

Nicolaas Hartsoeker also found it difficult to match the standard of Antoni's work. He was able to examine the 'little seed animals' of people, dogs and cockerels. But Antoni had even removed the testicles of a male flea and examined the seed animals they contained. Hartsoeker could not believe that this was possible. He made that clear during an angry visit to Het Gouden Hoofd – at least that was how he described it many years later.[11]

In his eyes, even the finest and sharpest knife would be unable to make a clean cut in the small blood vessels in these minute creatures, and would simply crush them, while the juices released in doing so would make a mess of the whole thing. Moreover, even with immensely magnifying lenses, it would be impossible to discern exactly where the knife was cutting. To see them properly, Antoni would have to hold his mites and fleas so close to the lens that when he inserted his knife between them, he would no longer be able to see the creatures and would have to cut blind. Finally, such a small piece of the specimen was in sharp focus that the knife would soon move into the vaguer part of the image. In Hartsoeker's opinion, all these objections made it impossible to prepare and examine the specimens in this way. So what tools did Antoni use, he asked?[12]

These were perfectly justified questions, but Antoni had no wish to speak to his angry visitor. According to Hartsoeker, his host told him only that he had his own unique type of lenses that he showed solely to Cornelia and Maria.[13] Antoni also revealed to the Englishmen Thomas Molyneux and John Locke in 1685 that he had his own special sort of private lenses.[14] And, according to Von Uffenbach, Antoni said years later that he had developed his own technique for producing special lenses that were not round.[15] Such claims arouse the curiosity. Did Antoni indeed have a different method of producing lenses than his colleagues? Is that why his work was so good?

Grinding, melting, blowing

Many lens-makers in Antoni's time ground the glass, producing lentil-shaped lenses with two concave sides, both of which formed part of a sphere. From 1658, Johannes Hudde – whom we briefly encountered earlier because he could make very small lenses – employed a different method.[16] He melted glass around the point of a needle, creating a small bead.[17] Robert Hooke did something similar when experimenting with small lenses for simple microscopes, but without using a needle. In his *Micrographia*, he explained how he melted a small piece of glass and extracted a thread from it. He then held the thread in a flame, producing a small bead with a piece of glass thread still attached to it.[18] This was the method that Hartsoeker learned from his mathematics teacher.[19]

Antoni used lenses made by grinding and melting, but possibly also devised another method.[20] In the twentieth century, lens expert

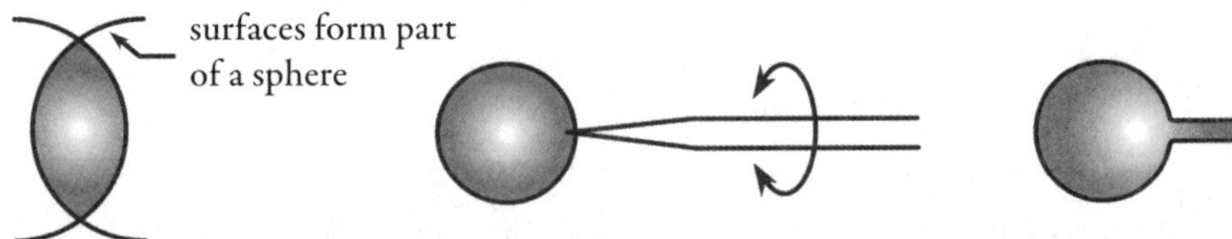

31 Grinding produced lentil-shaped lenses (1). Johannes Hudde made spherical lenses by melting glass on the point of a needle (2). Robert Hooke held a glass thread in a flame to create a glass bead at one end (3).

Jan van Zuylen suggested that Antoni may have blown (or had someone else blow) a glass ball in such a way that a lentil-shaped bead was formed directly opposite the opening. He could then detach the bead and use it as a lens.[21] If Antoni did use this method, he kept it very secret. There is no reference to it in any of his letters or other writings. And, in his time, that was both customary and not done: customary in the sense that the inventors or proprietors of useful techniques or lucrative knowledge were keen to keep their competitors in ignorance. In that way, they hoped to stay ahead of the rest.

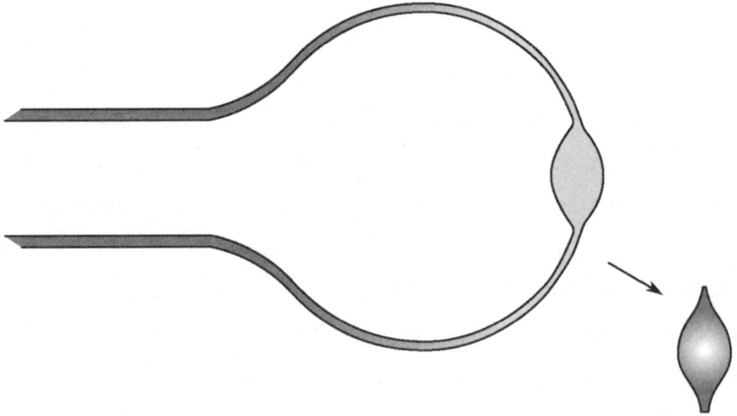

32 Proposal for Antoni's secret method for making lenses: blow a glass ball to create a lentil-shaped bead that can be used as a lens.

That's why trading enterprises like the Dutch East India Company (VOC), which had a branch in Antoni's home town of Delft, tried to safeguard their knowledge of coastlines and sea routes, and why craft guilds wanted to keep their production methods secret.[22] Not that these efforts were successful for long, as shared secrets tend to quickly end up public knowledge. But the intention was definitely there.[23]

From that perspective, it is logical that Antoni kept such a technical breakthrough to himself and did not share it with his outraged competitor, Hartsoeker. He may not have earned any money with his lenses, but they brought him status and esteem. And, in that respect, Hartsoeker had shown himself a rival by wanting to take the honour of discovering the little seed animals. Why would Antoni share his innovations with his opponent?

On the other hand, there was the etiquette of the scholarly world, of the Republic of Letters. The Republic was an international network of knowledge-lovers, who shared their ideas and insights in letters, books and periodicals, and by visiting each other.[24] The network had an unwritten rule that its members helped each other without reservation, even at the expense of their own fame.[25]

Moreover, there were the requirements of the Royal Society and like-minded scientists, who insisted on new developments being verified before they were accepted as valid.[26] That principle applied to all branches of natural science, but especially to microscopy, where error was a constant danger. Air bubbles, dust particles and other imperfections in lenses could, together with human failings, create all kinds of illusions. That happened to Antoni himself when he observed non-existent spermatic vessels. Many of his contemporaries were extremely cautious, especially when microscopists made curious discoveries. The Royal Society advised Antoni not to allow himself to be misled by his own findings.[27] To persuade his readers that his globules and little animals were real, Antoni had to give them the opportunity to replicate his observations. And that meant informing them clearly on his way of working and his results.

Illustrations

He did that in all kinds of ways, especially when talking about his observations. Antoni went to great lengths to give his readers a clear picture of his unexpected, unprecedented discoveries and to prepare them for what they too would see if they tried to find their way in this unknown miniature world. That is why he wrote that the little animals reminded him of tortoises or magpies, and that certain items looked like parrots' beaks. He had a whole arsenal of comparisons designed to build a bridge between the new, tiny world and the more familiar big world: veins in cerebral membranes reminded him of the curled creepers of grapevines, globules of fat in milk of drops of mercury, particles in wine or beer casks and so on.[28]

That was useful, but the figures he sent along with his letters were much more direct, giving his readers the impression that they were looking over his shoulder. Yet it remained an impression, as the drawings were interpretations, far removed from what Antoni actually saw under his microscope. In reality, what he observed was much less clear than the drawings suggested. That was in the first instance related to a problem that all drawing artists encounter in practice, namely that reality is three-dimensional, more dynamic and complex than even the most talented artist can reproduce.[29] Moreover, the lenses – and especially the stronger ones – only gave a sharp image of minuscule details and not a clear picture of the

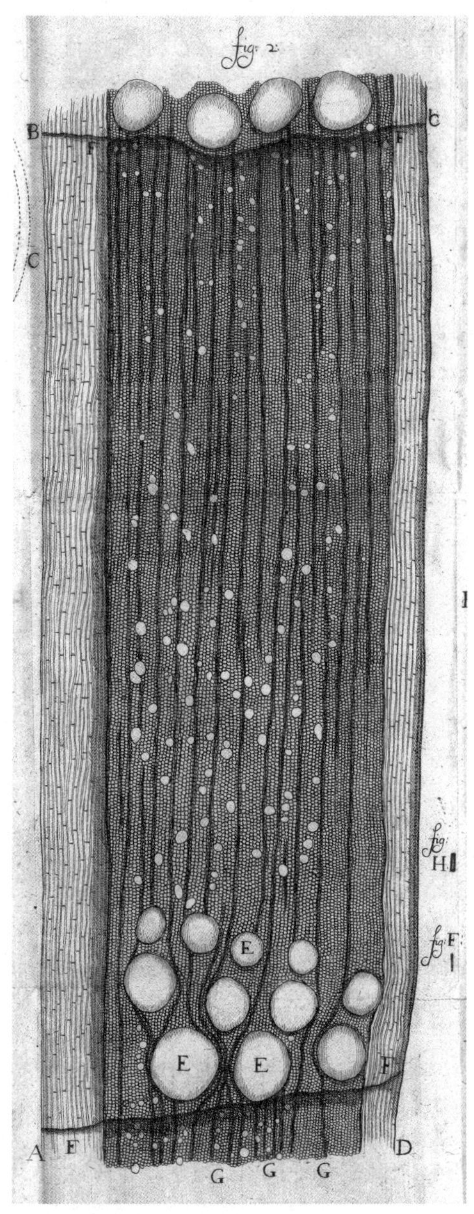

33 Although Antoni claimed that he could not draw, he made this illustration of oak wood.

whole. To achieve a comprehensible 'big picture', Antoni therefore had to 'paste' a number of observations together.

Although his stepfather and stepbrother were artists, Antoni had no training in the arts and believed himself to be bad at drawing.[30] He had a certain degree of talent, as can be seen from his very competent illustration of a cross-section of oak wood.[31] But drawing was not his forte, so he regularly hired professionals to depict his discoveries in the best possible light.[32] They looked through his microscopes and drew what they saw. Besides good illustrations, this also provided the added advantage that they were witness to his observations, making them a little more reliable.[33]

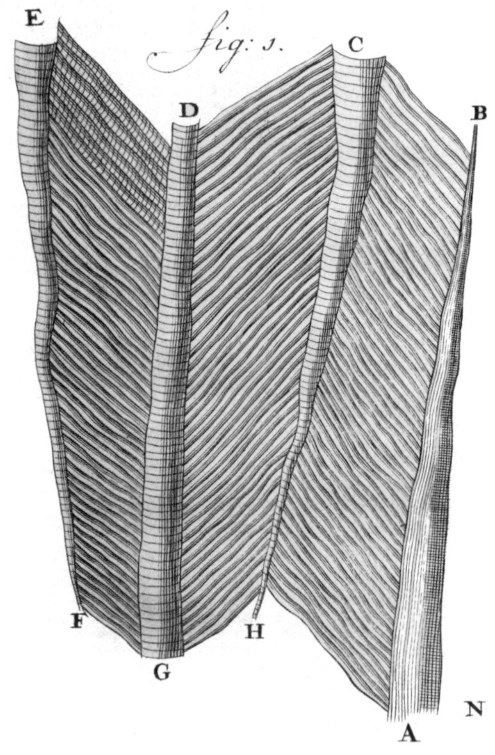

34 In Antoni's final years, Willem vander Wilt made his drawings, including this one, of the muscle fibres of a mouse.

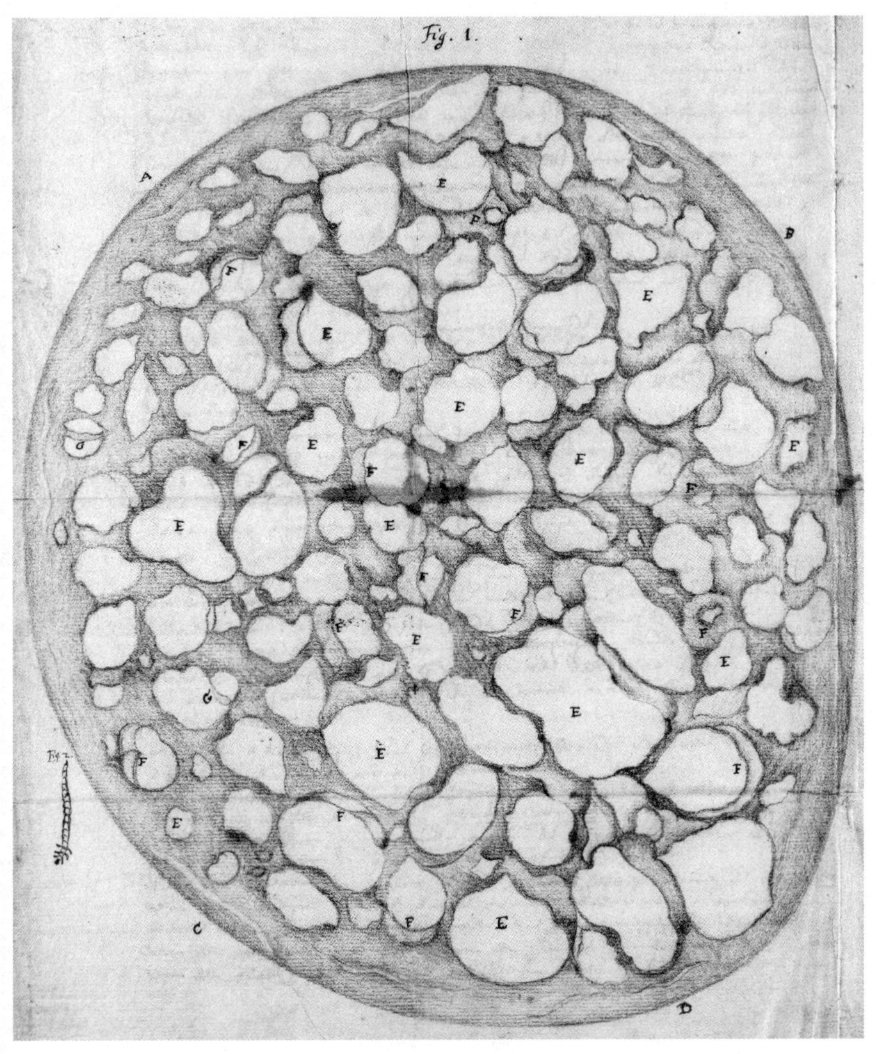

35 Antoni hired an anonymous artist to draw a cross-section of the optic nerve of a cow. Later, he also sent nerve samples to London.

We know the name of one of these artists, the last. He was called Willem vander Wilt, and his contemporary, Reinier Boitet, said of him that he could draw so well that 'few could equal him'.[34] From 1713, he regularly looked at Antoni's observations and drew what he saw, in consultation with Antoni.[35] When Antoni was over eighty and his own eyes failed him, he knew he could trust those of Vander Wilt.[36]

The descriptive texts and illustrations gave less talented microscopists a solid base from which to start, but that was not the whole story. To replicate Antoni's observations, it was essential that they made good specimens, and that was at least as difficult to learn as the new way of observing them. That was the reason why Hartsoeker was so keen to know what kind of knives Antoni used to prepare his specimens with such subtlety. There was no magical solution to that problem. It mainly required a lot of patience and practice.

Because it was so difficult to prepare his specimens, Antoni sent pre-prepared samples to London along with his illustrations. In the winter of 1674–5, for example, he sent a drawing of his observations of the optic nerve of a cow to the Royal Society. When the society doubted the verity of his findings, he sent a number of prepared nerve samples so that the curious scientists in London could see them for themselves.[37]

Fraternal advice

In addition, he gave his fellow scientists all kinds of practical tips and advice, as, for example, on the study of fleas. When he wanted to know how the insects were born, he collected their eggs. But he did not know when they would hatch. Because he could not afford to sit beside the eggs for days or months in his comptoir, he carried them around in his pocket. That meant he was there when the 'grubs' (larvae) emerged.[38] He had similar advice for studying silkworms. He also carried their eggs with him, but now to keep them warm. As the worms originated from warmer climates, Antoni wrote to the Royal Society, he even took them to bed with him and asked Cornelia, who clothed herself 'very warmly', to carry them 'in her bosom night and day'.[39]

He adopted a comparable method for the study of lice, and his explanation illustrates the social relations typical of his time. He wrote to a friend that he wanted to see how many eggs lice produce by giving new stockings to an orphan boy. He was planning to put two or three adult female lice, which were probably carrying eggs, into the stockings. He would then tie the stockings above the knee in such a way that no lice would be able to get in or out. After six or seven days, he would look at how many lice were living on the boy's legs. But he changed his mind, believing that he would 'gain greater certainty' by performing the experiment on one of his own legs. Moreover, he thought it perhaps fairer not to spare himself 'the itching and pain on one of my legs which most poor people are obliged to bear their whole life all over their body'.[40] So he shut two lice up in his own stockings and observed after six days that one had laid 50 eggs and the other 39 or 40. He also discovered one young louse.[41] He left the young louse on his leg to give it a chance to grow further, together with a number of unhatched eggs. A few days later, the louse proved to have grown considerably, and had been joined by 25 new young lice, while many of the remaining eggs were about to come out. That was too much for Antoni, who was suddenly 'filled with such aversion to the stocking' that he threw it, along with all the lice, out of the window.[42]

Antoni also used his own body for research on other occasions, sometimes in a very strange way. One time, for example, he wanted to know if maggots could live on earwax. To do that, he collected wax from his own ears for weeks and kept it in a glass tube. He then placed a number of freshly hatched maggots in the tube. Though there was nothing to eat other than the earwax, the maggots refused to touch it, even after Antoni had made it a little moist to make it more digestible. After a few days without food, a few of the maggots died and the rest looked soon to follow. So Antoni decided to add some meat to the earwax. The maggots ate the meat and even started to grow. But once it was gone, they again refused to touch the earwax, despite it having absorbed some of the tasty meat juices. So Antoni added a second piece of meat to the earwax and, sure enough, once they had devoured the meat, the maggots did eat the wax. Antoni concluded from this experiment that earwax was not suitable food for newly born maggots but was acceptable to slightly larger ones.[43] And, thanks to his clear explanation, everyone could check his findings.

Antoni also provided more technical instructions, for example on his study of blood. To do that, he tied off the tip of his thumb, the way surgeons did with arms if they wanted to let blood. He then pricked his thumb with a needle, releasing a little blood. He wiped that away, wanting blood that flowed directly from the body with no contact with air. So he took a thin glass tube with the diameter of a human hair and applied it to the small hole in his thumb. He then squeezed the tip of his thumb so that blood flowed into the tube. When he had enough blood, he laid the tube on a piece of white paper and broke a small piece of it off with his nail. He stuck that with a little saliva or turpentine to the needle of his microscope, or placed it in a copper tube attached to the microscope.[44] That enabled him to examine the blood clearly. He also varied the size of the tubes. If he wanted to observe the red globules, a thinner tube was better. But if he wanted to see the fluid in which they moved, or how they subsided in stagnant blood, the tubes could be a little thicker.[45]

These tubes were useful for all kinds of other experiments, partly because their curved shape meant that they also worked a little like a magnifying glass.[46] Antoni used them quite frequently, for example when examining the brains of a cow. As he explained to his readers in London, he first used a small knife to make an incision in the tissue. He then took a tube with the diameter of a coarse horse hair. He placed one end of the tube in the white part of the brain, moving it in and out while sucking on the other end. This filled the tube with pieces of tissue that he placed, still in the tube, in front of a lens.[47] He described this all very clearly, and was even thoughtful enough to send a few of the glass tubes to London, so that his colleagues at the Royal Society could use them themselves.[48]

Amid all this generosity in his contact with the Royal Society, Antoni occasionally showed a different side of himself, especially in the beginning. He wrote in 1675, for example, that he had developed a new method of observation that was so reliable that, if he were to make it known, everyone would admire him.[49] But he had no intention of doing that. Shortly afterwards, he wrote – arousing the curiosity as much as his earlier claim – that he had developed his own method for studying salt but that he did not wish to share it with anybody.[50]

Antoni thus conducted himself from time to time just as obnoxiously as Hartsoeker alleged. But much more often, he was helpful and free with his advice, including on how he used his microscopes. Point the instrument at the light, he recommended, as if it were a telescope with which you want to look at the stars, but during the day.[51] Or use a candle for extra light, or a concave metal mirror to direct the light. But do not look directly at bright light, like that of the sun, because then you will see all the colours of the rainbow, which are not there in reality.[52] A dark room, with only a little soft light from a candle, can also work well.[53]

A final category of wise lessons was about preparing specimens. Do you want to see little seed animals clearly? Mix a pinhead of seed animals in a drop of clean rainwater and spread it out over a flat, clear glass slide. In that way, you will distribute the animals more widely and be able to see better how they swim around.[54] Are you examining the thread of a silkworm and you can't see its structure clearly? Curl it up a little, and then you will see that it is actually a double thread, and that both parts are flat.[55] Are you studying the structure of beef, but your sample is almost invisibly fine and thin, even under a microscope? Colour the meat particles with brandy with a little saffron mixed into it. Then you will see them better.[56]

This sort of advice on how to prepare little animals, fluids and other items may have been the most important tips that Antoni could give. Contrary to what Hartsoeker believed, his research was not about the sharpest lenses. For a lot of his discoveries, he actually used less sharp lenses, because they gave a greater field of vision, giving him a better overview and enabling him to better understand what he saw.[57]

To achieve that, it was essential that the little animals, fluids, pieces of wood and so on were carefully prepared. That became clear to the team of researchers at Visualizing the Unknown when they replicated Antoni's work: without good preparation, they got nowhere. Fortunately, two of them were experienced biologists with three centuries of biological knowledge that did not yet exist in

Antoni's time, and they were equipped with modern microscopes. But, for them too, the preparation work was a big challenge. It must have been even more difficult for Antoni, who had to discover so much for himself and did not know what he was looking for. It is therefore not surprising that he was occasionally mistaken in what he saw, and that is why he gave this wise advice to the Royal Society: 'a person who is trying to publish new discoveries should not judge by one look, but he should see them many times.'[58]

Scanning lenses

All in all, thus, Antoni's desire to help out outweighed his pride. So why was he so secretive towards Hartsoeker and others about his lenses – assuming that his account was based on truth? And what method was he referring to in his conversation with Von Uffenbach?

Had it perhaps something to do with three striking items in the catalogue of Antoni's microscopes, which were sold after Maria's death? The lenses of these microscopes were not made of glass but of 'Amersfoort diamond', a quartz crystal.[59] And there was also one lens made from a grain of sand.[60] That may have been different to the more customary method, but it was not Antoni's secret. The rest of the few hundred lenses were simply made of glass.

Written descriptions therefore bring us no further. That's why the Rijksmuseum Boerhaave and Delft University of Technology tried a different approach. They decided to look at the lenses themselves. But that was no simple task, as the lenses are clamped, partially covered, between the two metal plates of the microscopes. And the microscopes are now so rare that researchers may not open them for any reason. They are far too valuable for that to be allowed. Fortunately, today we have neutron tomography, a technique that allowed the researchers to examine the lenses without damaging anything.[61]

Two microscopes were examined using this method: one from Boerhaave, which magnified 118 times, and the strongest surviving microscope, from the Utrecht University Museum, which magnified 266 times and was of special interest because it had caused lens expert Jan van Zuylen to speculate on how it was made: he had seen that the lens was exceptionally smooth on the outside, suggesting to him that it had been produced by blowing.[62]

It was fascinating research, but the result was surprisingly commonplace. The lens from Boerhaave proved to have been ground in the normal fashion, and the one from Utrecht had been made using the same method employed by Hooke and Hartsoeker. As the researchers noted in 2021, there was still a fragment of glass thread attached to the lens, from which it had been made.[63]

These are, of course, only two microscopes of the hundreds that Antoni made. So the results may be coincidental and it is possible that Antoni used a unique method in other cases. But Van Zuylen's theory about the Utrecht microscope is in any case incorrect. And perhaps Antoni deliberately misled his visitors, especially his competitor Hartsoeker and his arrogant guest Von Uffenbach.

This was not uncommon among craftsmen in Antoni's time. I explained above how enterprises and guilds kept their trade secrets safe from their competitors. Although their efforts were often in vain and the secrets leaked out, they often persisted in pretending that they were ahead of the pack, as if they were the only experts. That was good for their reputations.[64] It is possible that Antoni also did something similar, to make a good impression: especially in his early years with the Royal Society, and later with critics like Hartsoeker or young know-it-alls like Von Uffenbach. That could also explain something that he wrote to the Royal Society in 1699. He was certain, he told his readers, that the researchers there had excellent lenses, as he had exactly the same type of lenses himself.[65]

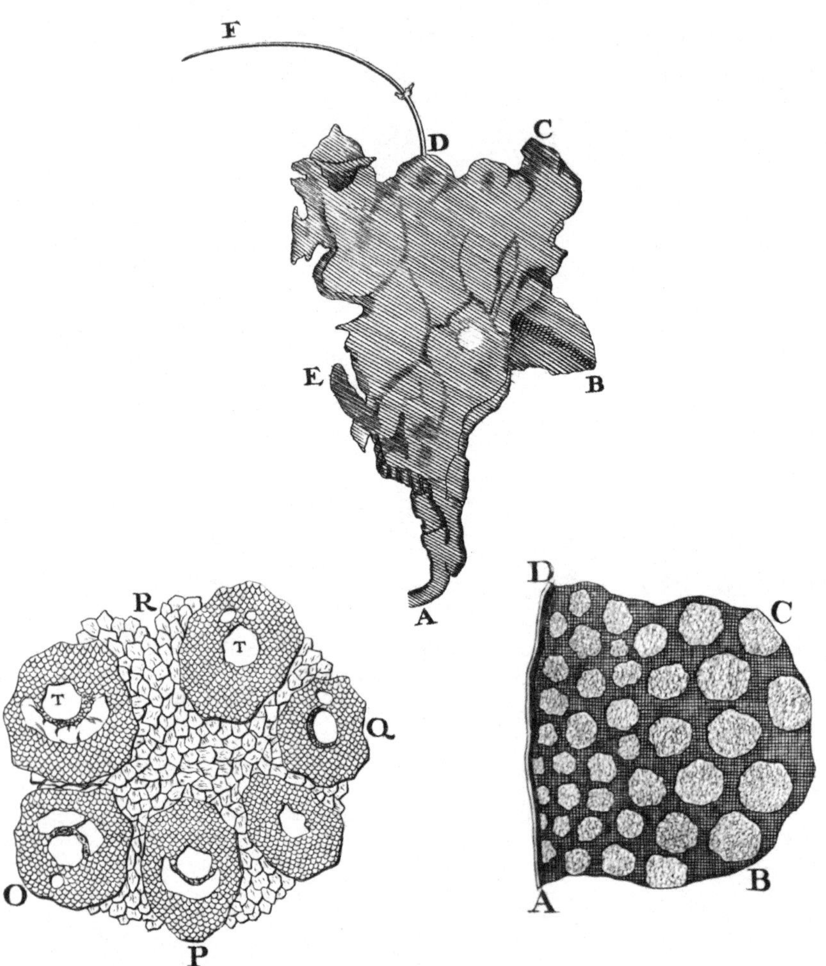
F
D
C
E
B
A
R
T
T
Q
O
P
D
C
B
A

SEVEN

Strange Curiosities and Senseless Details

We do not know her name, but she was about thirteen years old and lived in Delft with no parents. Her ancestors came from southwest Africa, where Dutch merchants bought prisoners and shipped them to the other side of the Atlantic Ocean as slaves. The girl's grandmother lived as a slave in Brazil, where the Dutch had a colony from 1630.[1] Perhaps she worked on one of the many sugar plantations.

When the Dutch lost the Brazilian colony in 1654, it is not clear what happened to the grandmother. But in 1684, her former owner lived in Delft. It is therefore probable that her granddaughter was brought to the Netherlands by the people her grandmother had worked for.[2] The granddaughter was officially free, as slavery was not permitted on Dutch soil. But the question remains to what extent she was able to control her own life.[3]

For Antoni, this 'Mooress' – as he called her – offered an opportunity to conduct research. He wanted to know why some people had black skin. In his time, there was a theory that Africans were by nature pink, just like Europeans. The black colour was supposedly caused by them rubbing a certain kind of oil into their skin. But Antoni thought that implausible.[4] He knew that human skin continually sheds flakes and concluded that dyed skin would thus gradually lose its colour. As that did not happen, there must be some other mechanism at work. To investigate that, he wanted to examine the girl's skin.

Using a 'fine little instrument', he took small specimens from the outer skin of the girl's arms. The skin beneath proved to be just as dark – an additional indication that the girl was not dyed but was dark-coloured by nature. Antoni also tried to take some specimens from the underlying skin, but – in spite of the 'care he took' – the girl began to bleed a little, which 'hindered him in his work'.[5]

When he examined the 'scales' of skin through a microscope, they were transparent and 'showed slightly black'. They were not as black as the whole skin, but Antoni concluded that was because the scales were very thin. Multiple layers on top of one another would make the skin appear black. Thus it was clear that the black colouring came from inside.[6]

Showcasing exotic curiosities

The girl fascinated Antoni because she was different from the other people of Delft, and because he was always attracted by the strange and exotic. That was also clear to see in the specimens he left behind, together with his microscopes, and which were sold after Maria's death. One was a piece of 'Silver from the mines, in the form of a small man'.[7] And, all the way from the Moluccan Banda Islands was the 'Little animal that eats nutmeg'.[8] Equally exotic were the 'Thorns of the Indian Fig', which we would now call the spines of the cactus fig, which originally came from America. The plant owed its name to the Spaniards, who had once mistaken the American coast for India.[9] Finally, from a little closer to home, were the hairs of an elk.[10]

Antoni's fascination with the exotic was typical of his time. Wealthy seventeenth-century Dutch citizens liked to fill their homes with curious or rare objects.[11] They collected dead animals, Asian latticework, antique coins, the horns of unicorns (that is, narwhal tusks) and other such strange and wonderful things. Antoni's collection of microscopes with specimens of the natural world was a variant on this preoccupation.[12]

Antoni obtained some of his curiosa from seamen who sailed to Asia with the Dutch East India Company (VOC). He once asked them to bring back a piece of a coconut palm, so that he could examine it. That was not an immediate success, as they ignored his request and came home empty-handed. But in 1716, a helmsman brought him a piece of palm tree from Java. It had suffered considerably from the voyage and had gone partly mouldy, but Antoni was still able to obtain a reasonably good impression of its structure.[13]

Sometimes, he would come across interesting items completely by coincidence. One day, for example, he saw a furniture-maker carrying wood into a neighbour's house. The wood was red, with white and black flecks, and came from the Moluccan island of Ambon. The furniture-maker had bought it from the VOC. Antoni wanted to know where the colours came from and asked for a few fragments to study. When he put them under a microscope, he saw that there were large veins running through the wood. Where the wood was red, the veins were filled with red fluid.[14]

He also obtained research materials from friends who were interested in research. Constantijn Huygens Junior, for example, gave him moxa, a herb used in Asia to treat gout and other complaints.[15] Antoni did not suffer from gout, but he was interested to know more about the herb, which he understood you had to burn rather than ingest. So he laid a little on the back of his hand and set fire to it. That resulted in a nasty burn. It is not advisable to use moxa directly on the skin (there should be some distance between the burning herb and the patient).

After the burn, he stopped experimenting with the moxa. Not because of the pain, he wrote stoically; he could tolerate that. The wound was no worse than a knife cut in his hand, which he would simply stitch up himself. But it healed slowly, which hindered him in his work.[16] So he examined the moxa through a microscope and concluded that it was not a herb, as had been claimed, but the 'expiration' of a fruit, like the down on peach or the lint of another exotic material, cotton.[17]

A somewhat chaotic type

Studying such curiosities was fascinating, but in Antoni's time, interest among collectors began to shift. Collecting curiosa gradually made way for more specialized collections of, for example, minerals, seeds or old coins. The focus moved away from the exotic nature of the items and towards systematic categorization.[18] Behind this new type of collection lay an ideal of knowledge grounded on order and classification: plants, animals, rocks and so on had to be categorized in species, genders and classes. They could then be compared methodically, so that we humans could better understand the whole natural world. According to that ideal, real knowledge was based on systematic analysis. At first glance, Antoni's writings seemed not to fulfil that criterion. Letter after letter contained a mixture of observations, but he never provided a systematic clarification or a synthesis of his ideas and findings.

Antoni's self-proclaimed enemy, Hartsoeker, took pleasure in this, portraying Antoni as a somewhat chaotic type who filled his letters with badly organized ideas.[19] He claimed that Antoni was a capable artisan who repeatedly used the same trick with his lenses, but never came any further than senseless details.[20]

That was doing Antoni a misjustice. He conducted a great deal of comparative research, also with his exotic specimens. For example, he compared the 'scales' (skin cells) of the thirteen-year-old girl with his own, which proved to be bigger and more transparent than the former's.[21] And he compared the moxa with cotton, to which it was similar in appearance. He made small piles of both and set fire to them – not this time on his skin, but on a pile of paper. That produced two identical flames, leading Antoni to conclude that treating a gout patient with burning cotton would have the same effect as with moxa. After all, the crucial factor was the fire – he saw no other option – and that was the same in both cases.[22] These comparisons may appear rather superficial, but this was completely new terrain and Antoni made serious attempts to understand it.

In many cases, Antoni spread his comparisons out over a number of years. He would examine a different kind of salt, blood or plant seed, and use his findings to build on earlier observations. And every time he had something new to report, he wrote a new letter.[23] Unfortunately for his readers, he often failed to provide a summary of his earlier ideas, so that – to achieve a clear picture – they had to read back through his old letters. But anyone who took the trouble to do that would see that he produced much more than senseless details.

That was also true of his study of wood. As early as 1673, Antoni described the structure of different kinds of wood, including pine and oak, and the 'pipes' he saw running through them. In 1721, almost half a century later, he repeated his observations.[24] And in the intervening years, he had studied many kinds of wood, including the Javanese palm and the planks from Ambon. Thus, this exotica too fitted in with his long-running research programme.[25]

During this ongoing research, Antoni regularly corrected himself if he discovered that he had interpreted things wrongly in the past. A good example of that was the spermatic vessels that proved not to exist, and the painfully sharp 'small rods' that he had once seen in eel's blood and that later turned out to be round discs. And he did that even if it eroded his own fundamental ideas, such as his basic assumption that all animals had more or less the same proportions. That principle did not apply to the little seed animals, as he discovered when he examined a flea. The flea's seed animals were rather large for such a small creature: proportionately bigger, for example, than those of a human.[26]

Born from mud

So Antoni was clearly analytical in his approach. Moreover, his research very much addressed scientific discussions of his time. One of those played a prominent role in his work: the question whether animals could generate spontaneously, without the need for parents. The idea that this was possible had been around for at least 2,000 years, as we can see from Aristotle's writings in the fourth century BC. Aristotle believed that eels, molluscs and insects could form in media such as mud, rotting slime on a ship's hull and meat.[27]

There were a number of variants of this theory in the seventeenth century. One of the most influential, proposed in the 1660s by Jesuit and polymath Athanasius Kircher (1602–1680), was based on the notion of a ubiquitous universal seed known as *panspermia*.[28] These terms perhaps call to mind the seeds of plants, or Antoni's 'little seed animals', but Kircher meant something of a completely different order. His universal seed had been made by God during the Creation, of the finest imaginable material. It permeated all things, plants and animals, and ensured that they functioned as they should.[29] When creatures died, their seed would be spread around the world by wind and water, but some residue of the seed would always be left behind in the material remains of a living being. That residue could generate new life, albeit from a lower order than the original creature. Thus could insects be formed from rotting flesh and fungi grow on dead trees.[30] This new life was not generated entirely spontaneously, but it did not require parents to mate.[31]

A contemporary of Kircher, Francesco Redi (1626–1697), tested this idea by placing thousands of pieces of plant, meat or fish in jars. He sealed some of the jars and left the rest open, and then he waited. Maggots soon appeared in many of the open jars of meat and fish, but not in the sealed jars. He concluded in 1668 that this was due to flies being attracted to the food. But they could only get into the open jars, where they laid their eggs. And that is where the maggots came from, and not from Kircher's universal seeds. Flies were therefore required to make new maggots and flies.[32]

*

Researchers in the Dutch Republic also addressed this question. One of them was Jan Swammerdam, the other talented microscopist and contemporary of Antoni. Like Kircher, he based his theory of spontaneous generation on his religious beliefs, but he came to a different conclusion. According to Swammerdam, God had created all life on the basis of the same principles, and that also applied to reproduction. It would be very strange theologically speaking, he reasoned, if some animals were exempt from the

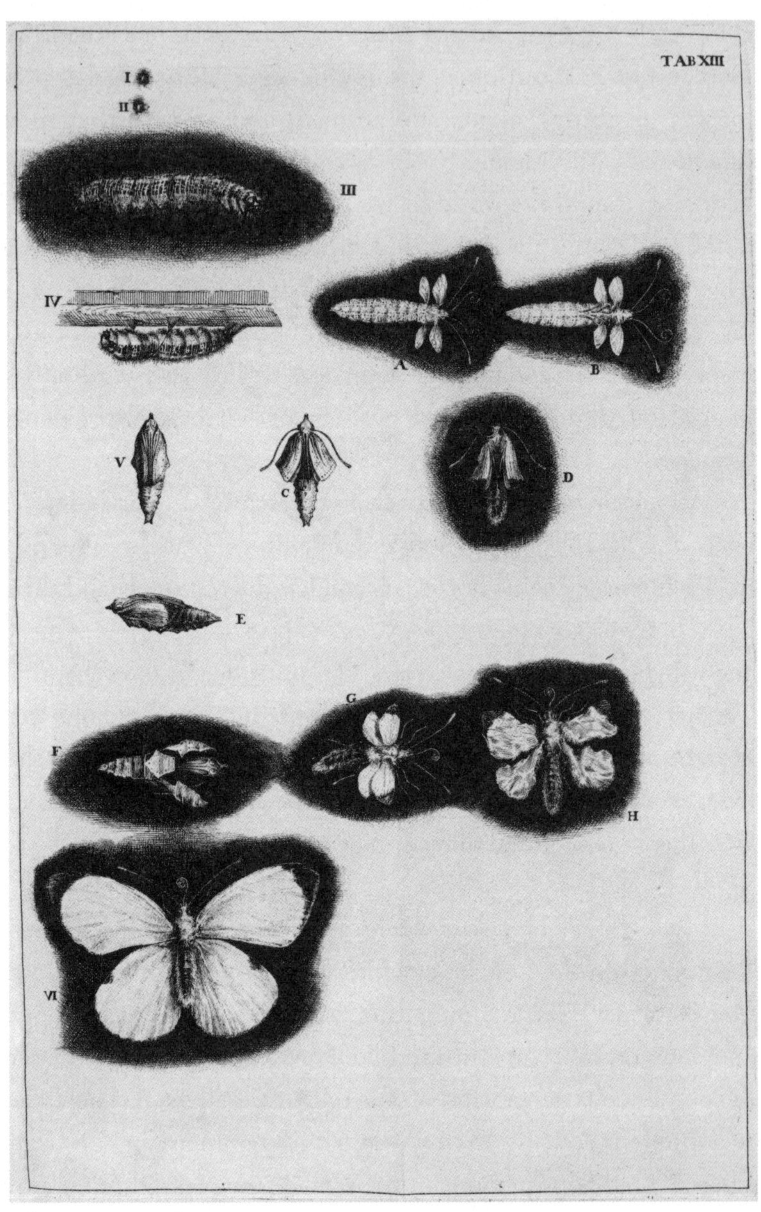

36 Jan Swammerdam observed how a caterpillar metamorphosed into a butterfly.

standard procedure and could be generated spontaneously from mud or rotting flesh.[33]

From the late 1660s, Swammerdam conducted comparative research into a wide variety of animals and how they reproduced. Together with Leiden professor Jan van Horne, he examined the 'eggs' in the 'female testicles' and became convinced that these eggs were crucial for all animals. In addition, he investigated metamorphoses, for example of caterpillars to butterflies. A variant of the spontaneous generation theory claimed that the original animal (for example, the caterpillar) died during the metamorphosis and a new animal (the butterfly) emerged from the rotting remains. Swammerdam showed that this was incorrect, and that the components of the butterfly were present in the caterpillar.[34]

While Swammerdam was working on his groundbreaking discoveries, Antoni was also becoming acquainted with microscopy, and they heard of each other's work. In theory, that could have been beneficial for both of them, as they had much in common: they were of the same generation, were both influenced by the mechanistic particle theory, and both had a great deal of talent for working with a microscope. But things turned out differently: in 1674, Swammerdam visited Antoni twice in quick succession, but they did not get along.[35] Antoni found Swammerdam not very cooperative: he listened to all the information that Antoni freely shared with him – in Antoni's words – but gave nothing in return.[36] Years later, in 1694, he would even blame Swammerdam for the death of Regnier de Graaf. The two had argued vociferously about who had first observed the female testicles. According to Antoni, the argument had affected De Graaf so deeply that he fell ill and later died.[37] For his part, Swammerdam felt that it was impossible to hold a decent conversation with Antoni, because he had not enjoyed an academic education and was severely prejudiced.[38]

So their personalities clashed but, like Swammerdam, Antoni refuted the theory of spontaneous generation. It is unclear exactly when he became convinced that the theory was nonsense, but when he saw in 1680 that even the tiny flea produced seed animals, he wrote it down clearly and unambiguously: all animals were created by reproduction.[39] Antoni would seek conclusive proof of this theory for many decades, long after Swammerdam's premature death in 1680. To do that, like Swammerdam, he observed many different kinds of insects, all of which proved to reproduce in the 'normal' way.[40] Furthermore, he once again saw how complex and refined the creatures were. He could not believe that such a complex animal could be generated spontaneously from mud or rotting flesh.[41]

All other kinds of other fauna and flora that Antoni studied proved to reproduce. He believed that he even had evidence to prove that this was the case with molluscs, which Aristotle had claimed could spontaneously appear on ship's hulls. In 1694, he examined mussels from Zeeland, searching for seed animals and ovaries, as he believed that the eggs did have some significance in the reproduction of non-mammals. He did not find what he was looking for, but he did see fine structures on the shells, which he took to be clusters of minuscule eggs. He suspected that here a new generation of mussels was evolving. That was incorrect, as young mussels do not adhere to the shells of their parents. What he saw were probably colonies of *Membranipora membranacea*, a common species of small creatures that form colonies in seawater.[42] But for Antoni, this was a convincing addition to a long list of instances disproving spontaneous generation. He was firmly convinced that 'just as it is impossible for a whale to spring from mud, it is also impossible for a little shell-fish to arise without generation.'[43]

That was a clear summary of his findings, but for many of his studies Antoni did not describe his conclusions and his readers had to construct the bigger picture for themselves. He received sufficient encouragement to share his research more freely with others: critically from Hartsoeker, and in a more friendly tone from German philosopher Gottfried Wilhelm Leibniz (1646–1716). Leibniz was fascinated by Antoni's work and hoped that, with the city's support, he would lead a school for microscopy in Delft.[44] But, as Antoni wrote to him in 1712, there was no chance of that, as he did not wish to have obligations to students. He was too fond of his freedom.[45]

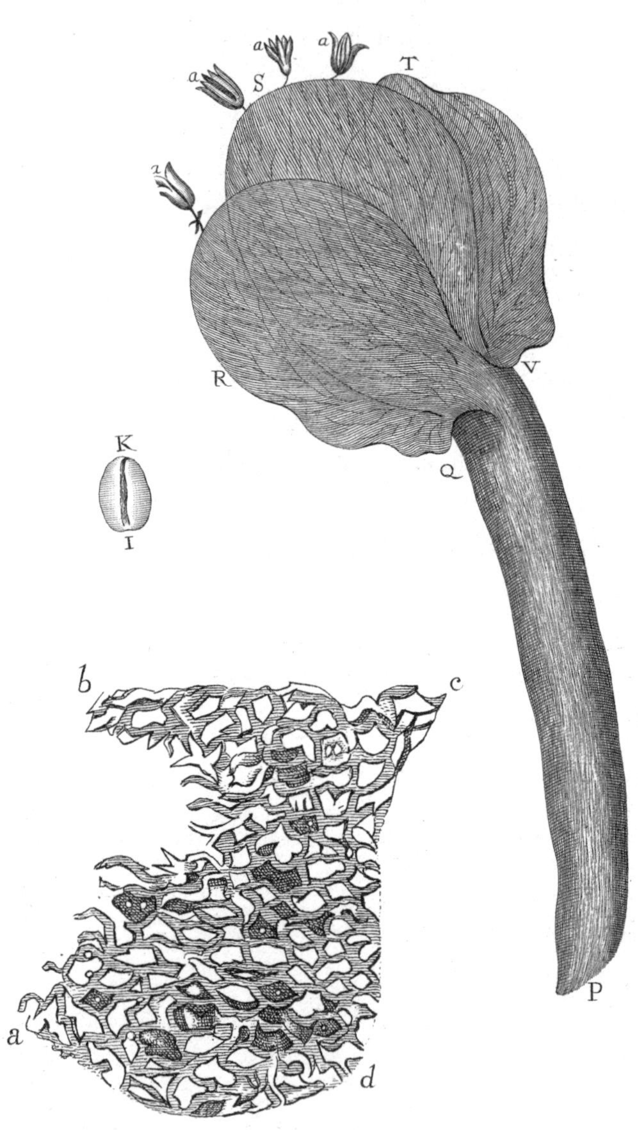
a
a
a
S
T
a
R
V
K
Q
I
b
c
P
a
d

EIGHT

Busy Times in Het Gouden Hoofd

Would Antoni be so kind as to accompany them to speak with Tsar Peter the Great? It was the autumn of 1697 and the Russian ruler was in Holland, to conduct politics and to expand his knowledge.[1] From The Hague, he had travelled by water to visit an arsenal just outside Delft. As the leader of an emerging power, he was very curious about military affairs.[2] But, besides war, he had a great interest in the mysteries of nature. So while he was taking a look at the arsenal, two of his men went to Het Gouden Hoofd to fetch Antoni. Now 65 years old, Antoni had been working with his microscopes for some thirty years. He had become a well-known figure in Holland and far beyond, and even attracted the attention of monarchs.[3]

For the tsar, Antoni took with him a number of microscopes and an *aalkijker* (eel-viewer), with which you could examine the blood circulation system in the tail of a fish.[4] As described in Chapter One, Antoni used one of these instruments in 1710 to show his young German visitor Von Uffenbach a fishtail. But that was a newer version, which Antoni had developed a few years earlier.[5] It is thus probable that, in 1697, the tsar had to make do with an old model. It consisted of a metal holder, a glass tube in which an eel could be placed and a lens held against the tube.

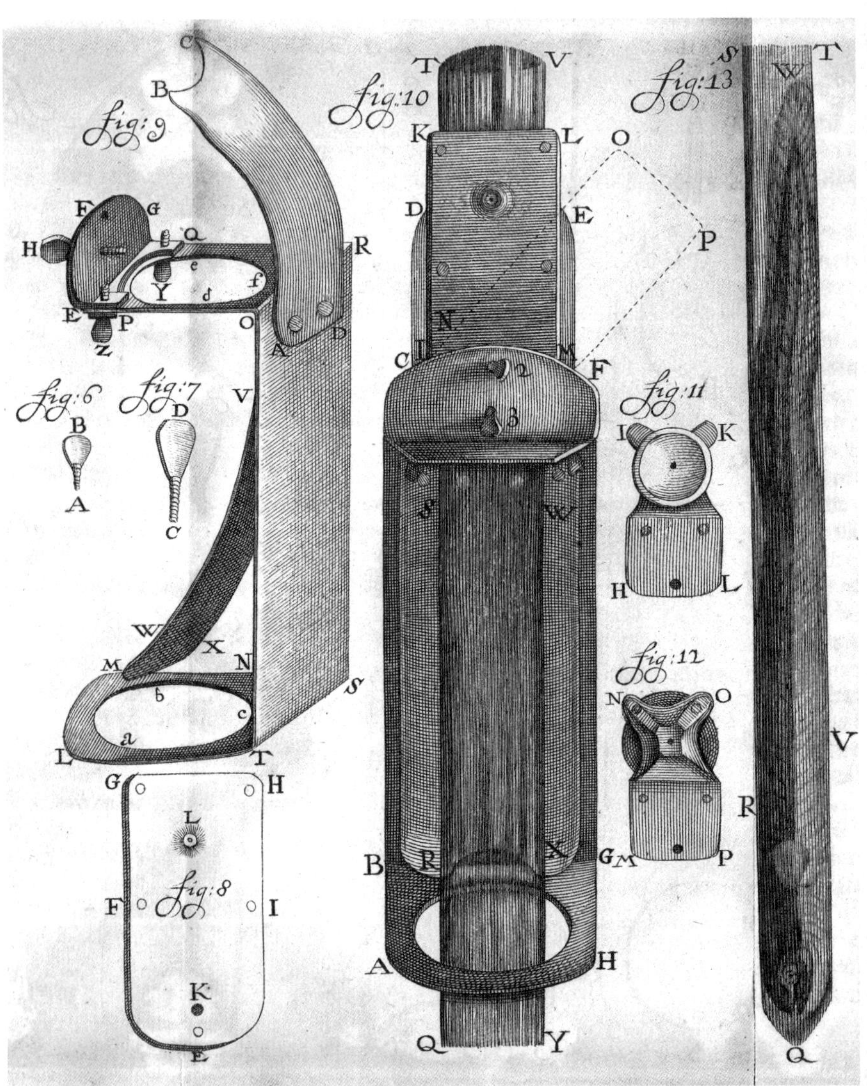

37 Antoni's first eel-viewer consisted of a metal holder (left), a glass tube holding a live eel (right; the eel was apparently so stupefied that it didn't move) and a lens in a metal plate (figure 8, below left), which was held against the tube. The whole instrument is shown in the centre. Antoni or his guests could use the viewer to see the blood flowing in the eel's tail.

Antoni took his instruments, and presumably a live eel, to the tsar, and – if we are to believe historian Gerard van Loon (1683–1758), who published an account of the meeting in 1731 – spent two hours delighting his royal admirer with the blood circulation in the eel's tail and other wondrous discoveries.[6] Van Loon is our only record of the meeting, as Antoni said nothing of it in his surviving letters. The encounter fitted in perfectly with Peter's itinerary during his visit to Holland, which provides some substance for the verity of the story. According to Van Loon, the meeting was a success. When they parted, the tsar expressed his great appreciation 'by shaking Antoni's hand'.[7]

Peter the Great was by no means the only prominent figure to show an interest in Antoni's work. In 1678, the English king Charles II had taken great pleasure in looking at little animals he had discovered in pepper water.[8] That discovery was described in Chapter Five, and was an unforeseen spin-off of Antoni's study of air and his attempts to explain the sharp taste of pepper. It happened in the spring of 1676, after he had seen the little animals in the Berkelse Meer, and had followed that up by studying rain and well water.[9] It was therefore no great surprise to find little animals in pepper water. But there were so many and such a wide variety that he reported extensively on his findings to Royal Society secretary Henry Oldenburg. Some were elongated, no thicker than the hair of a louse and with legs in front of their 'head', others were oval-shaped, or had tails, and there were many more. One sort was so small that 'if 100 of them lay [stretched out] one by another, they would not equal the length of a grain of course sand.'[10] In later letters, Antoni would inflate the number: in 1679, he claimed that three or four hundred of the smallest animalcules would fit into the length of a grain of sand. He used these figures to make his complicated sums on the little animals' capillaries.[11] But, whether a hundred or four hundred of the little animals fitted into the length of a grain of sand, they were certainly minuscule. And that is how modern-day biologists know that Antoni saw bacteria in the water.[12]

Antoni sent a detailed report of his observations of pepper water to Oldenburg, who published them in *Philosophical Transactions*.[13] But a substantial response from London was not forthcoming, so a year later Antoni returned to the matter.[14] This time he added a signed statement by witnesses who had seen how he had taken a small drop of pepper water, the size of a grain of barley, and 'deftly' sucked it into a thin glass tube. He had then divided the tube into fifty equal parts, which his spectators were permitted to examine through his microscope. They testified that, in each of the fiftieth parts, they saw at least two hundred living creatures. Two of the witnesses, Benedictus Haan and M. Henricus Cordes, both Lutheran pastors, then explained clearly how Antoni had determined the number of animals.[15] These testimonies were followed by others, signed by:

R. Gordon, medical student
J. Boogert, Licentiate of Civil and Canon Law, and notary public
Rob. Poitevin, Doctor of Medicine of the University of Montpellier
W.V. Burch, Licentiate of Civil and Canon Law, Advocate in the Court of Holland
Aldert Hodenpijl, manager of the Comanscolff, the St. Nicolaas guild house on the Ouden Langendijk in Delft
Alex. Petrie, Pastor of the English Congregation in Delft.[16]

The witnesses gave Antoni's observations added credibility. The more people who saw the same phenomena, the greater the chance that their observations were correct, and the smaller the chance of errors or optical illusions. This was also the thinking at the Royal Society: a single observation was no observation, and if several spectators confirmed a result, that made an experiment more reliable. And especially if the witnesses were men as, according to the norms of the time, with men the truth was in good hands.[17] Perhaps that is why Antoni chose pastors and others with the academic background that he lacked. The Royal Society appreciated that and considered the witnesses men 'of good credit'.[18] Two fellows of the Society announced, shortly after Antoni, that they had seen little insect-like animals in water that were invisible to the naked eye.[19]

That was all well and good, but it was not enough for the Royal Society. The fellows had not seen little animals in the spectacular quantities and of the minuscule sizes that Antoni described. And, rather than a series of testimonies on paper and the separate remarks of two fellows, what they wanted in London was to see the little animals with their own eyes. So they instructed Robert Hooke, with all his experience in microscopy, to replicate the observations with pepper water. His first attempts failed but, on 15 November 1677, he was able to show a group of fellows thin glass tubes filled with pepper water, in which thousands of little animals were swimming around. Consequently, on that day, the Royal Society decided that there was no longer any reason to doubt the verity of Mr van Leeuwenhoek's discovery.[20]

When the patron of the Society, King Charles II, heard about this he wanted to see the little animals for himself and was given a demonstration. The king was 'very well pleased', Hooke wrote to

Antoni, adding that he had of course mentioned the latter's name to His Majesty.[21]

Royal visit

According to Gerard van Loon, King Charles visited Antoni, as did kings Augustus of Poland, Frederick I of Prussia and George I, the last of whom became ruler of England a few decades after his visit. In addition, 'an over-abundant crowd of lesser princes, envoys and other prominent figures' paid visits to Het Gouden Hoofd.[22] Van Loon may have exaggerated a little here, as we have no other sources for the four royal visits than his own book. But that Antoni received considerable attention from the high nobility is certain. In 1679, for example, the Duke of York, later King James II of England, observed a louse and the seed animals of a dog.[23] James's daughter Mary also visited Antoni's house, just before she was crowned Queen of Great Britain in 1689. Unfortunately, Antoni had just left the city, so he was unable to give her a demonstration.[24] A little lower in rank was Elector Johann Wilhelm von Pfalz-Neuburg, who examined some of Antoni's discoveries in 1695, together with his wife and mother.[25] And the Prince of Liechtenstein visited Antoni in the winter of 1703–4 while he was in The Hague with the German emperor waiting for an opportunity to sail to Spain.[26] According to Antoni, the prince also tried to arrange a demonstration for the emperor but nothing came of it: before Antoni could travel to The Hague with his lenses, there was a favourable wind and the emperor was able to depart.[27]

Over the years, Antoni made a whole collection of microscopes with accompanying specimens, selecting the best lens for the creature, stone and so on that he wanted to observe. For a louse, for example, he chose a not overly strong lens, so that his visitors could clearly see the creature as a whole, and not just the details.[28]

Taking credit for Antoni's discoveries

Besides monarchs, Antoni also received visits from prominent figures from the Republic of Letters, as his reputation spread through the texts and conversations that formed the basis of that virtual land. Father and son Huygens were among the early visitors. Later there were others, including Leibniz, who visited Antoni in 1676 to see the spectacle of the little animals in pepper water, and John Locke, who came in 1685 and had some difficulty in seeing the tails of what Antoni suspected were little seed animals.[29]

A slightly less famed visitor was Cornelis Bontekoe (1647–1685), a follower of Descartes and 'tea doctor': he firmly believed that tea was a remedy for a whole range of complaints.[30] Antoni embraced the idea with verve, writing in his old age, in 1717, that he drank three cups of tea a day, plus two cups of coffee. He would double that dose if his urine was redder than normal, as he saw that as a warning that his blood was too thick, which was in turn a sign of illness. Hot tea and coffee helped to cure that by thinning the blood.[31]

Earlier, in the 1680s, Bontekoe was a regular visitor to Het Gouden Hoofd. Such visits from like-minded thinkers gave Antoni a valuable opportunity to exchange knowledge and ideas. They discussed, for example, the theory of Leiden professor Sylvius that food was digested as a result of 'effervescenses' produced when acidic and alkaline fluids reacted with each other. Antoni rejected Sylvius's way of thinking as too chemical for his view of the world. As he explained to Bontekoe, during the digestive process, globules were simply rearranged mechanistically. His visitor had listened attentively and announced that he would write on the issue, and would mention Antoni's name.[32] Bontekoe indeed published a pamphlet in 1682 criticizing Sylvius, as well as De Graaf's research on pancreatic fluid.[33] But he made no mention of Antoni's name.[34] Bontekoe had even included Antoni's ideas on the thicker blood of fever patients: because the blood does not flow as easily, the heart has to beat faster to pump it round the body. According to Antoni – and to Bontekoe – that explains why sick people with a fever often have a quicker heartbeat.[35]

If Antoni wished to be honoured for his work, there was clearly no point in waiting for others to write about him. Then there was a risk that, as he claimed Bontekoe had done, they would take credit for his discoveries. Antoni may have had a platform through the *Philosophical Transactions*, but that reached only a select, English-reading audience. Furthermore, it could take an age before his letters were published, if that happened at all. From 1686, and again from 1714, astronomer Edmond Halley (1656–1742) was in charge at the journal for a few years and Antoni found it almost impossible to be heard.[36] Halley preferred to publish on mathematics and physics than on the subjects of Antoni's research.[37]

In the meantime, a publisher from Leiden had opened up a new avenue. Daniel van Gaesbeeck had acquired two of Antoni's letters – to Royal Society secretary Thomas Gale and to Robert Hooke – and published them in 1684, in Dutch.[38] As usual with Antoni, the letters discussed a diversity of subjects, including the fermentation of beer, little animals in water and the reproduction of dragonflies.

Van Gaesbeeck had not asked Antoni's permission to publish the letters in advance, writing only in the introduction to the booklet that he hoped that Antoni would forgive him for his audacity.[39] That was indeed a little impertinent, even though ideas on intellectual property were considerably more relaxed in those times than they are today. However, a number of Antoni's acquaintances found Van Gaesbeeck's behaviour acceptable and encouraged him to give the publisher more letters. After sustained insistence on their part, Antoni consented, giving Van Gaesbeeck six letters in all. He published them in different combinations, resulting in seven partly overlapping booklets.[40]

That was as far as Antoni's collaboration with Van Gaesbeeck went. In 1685, he changed over to another publisher in Leiden, Cornelis Boutesteyn. They worked together for more than twenty years, bringing out a whole series of letters. Sometimes they would be single letters, but more often they were collections, which could be seen as Antoni's alternative for summaries of his work – but without the subject-matter being clearly ordered in any way. By far most of the letters were published in Dutch, though Boutesteyn did produce a small number in Latin translation for an international scholarly audience.[41]

The cooperation with Boutesteyn ensured that Antoni maintained control over the presentation of his own work, but did not mean that he was loyal to the publisher. From 1693, he also worked with Hendrik van Kroonevelt, who lived in Delft in the Hippolytusbuurt, in the corner house that has since been merged with Antoni's.[42] With his neighbour, Antoni published other letters than with Boutesteyn.[43]

Illustrious name

With his fourth publisher, Antoni pulled out all the stops. According to Van Uffenbach, Adriaan Beman – also from Delft – had the best bookshop in the city. Antoni started working with him after Van

38 The frontispiece illustration to Antoni's last collection of Dutch letters by Jan Goeree was full of references to antiquity.

Kroonevelt died.[44] In 1718, Beman published a book containing 46 letters, a lot for a collection of Antoni's texts, and shortly afterwards a Latin translation.[45] In the Dutch version, Beman devoted several pages to praising Antoni, starting with the cover illustration. It showed Antoni in his latter years looking kindly at a whole array of classical references to his achievements, accompanied by an angel heralding his honour on a trumpet.

The sitting woman wearing a floral wreath – symbol for the knowledge of nature – is lifting a sheet, revealing – like Antoni – things that had long been concealed. Beneath the sheet is a sphinx, which represents sharp-mindedness, with a book and a protractor to show that Antoni was a learned scholar. Everything in the illustration was wondrous: in the background on the left, ignorance (with ass's ears) flees the scene, as the accompanying description explains.[46]

The illustration is followed by poems by Arnold Hoogvliet and Hubert Kornelisz Poot. The two poets had also been called upon a few years earlier, when a group of professors from Leuven honoured Antoni with a specially minted coin and a book of texts praising his achievements as a great scholar and member of the Royal Society. The book began with a poem by Leuven professor J. G. Kerckherdere. As was customary, it was in Latin and thus not very suitable for Antoni. Fortunately, Hoogvliet had translated it into Dutch, and he and Poot had also written a number of poems themselves. Together, the poets waxed lyrical about Antoni's 'illustrious name' and praised the man who had risen far above his old source of inspiration, Descartes, had uncovered countless hidden secrets and had proven that Aristotle had been mistaken in his ranting about spontaneous generation.[47]

Hoogvliet and Poot wrote new material for the collection published by Beman, but there was some overlap. In both publications,

Antoni was compared to another scholar from Delft, Hugo de Groot (1583–1645), who had a great impact on the development of jurisprudence.[48] Antoni would not have minded this honourable comparison being repeated.

Antoni and Beman planned another collection of letters, but something went completely wrong. No one knows exactly what, but Antoni considered the issue serious enough to write in a will in 1721 that Beman was a scoundrel – though someone later crossed it out.[49] He also declared that Beman should no longer have anything to do with his letters. After Antoni's death, any unpublished letters were to be kept in a sealed chest, together with ten copper plates that had already been made for the printing of the new collection. As long as Beman was still alive, nothing more was to be published.[50] Those orders were carried out and, also after Beman died in 1736, the letters remained unpublished. Antoni himself had died, and no one else took the task upon themselves. Perhaps there were letters unknown to us, with more observations. Or perhaps they were letters that had been sent to the Royal Society or elsewhere, and which have been kept in the records, so that we know their content from other sources.[51]

*

Despite this setback, Antoni could look back in his old age on a substantial series of his own publications, for which there was considerable interest. He heard from an acquaintance in the Dutch East India Company (VOC), for example, that his letters were read as far away as Java.[52] Moreover, he had acquired the recognition of prominent figures, and from members of the Royal Society in particular. In 1677, he had still thought it necessary to have his discovery of little animals in pepper water endorsed by witnesses, but now he felt that the scholars in London knew how reliable he was.[53] He may have made the occasional mistake in the past, but generally speaking it had proven well worth taking his reports seriously – even the most implausible among them.

Recognition had also brought with it a great deal of interest, and so many visitors that it sometimes became too much for him. In 1711, he received no fewer than 26 visitors over a period of four days, which made him so tired that he 'broke out in a sweat all over'.[54] And they were only the people with letters of introduction from the right connections – which Von Uffenbach also had with him – plus an unnamed duke and count with his tutor.[55] Others, he – or more probably his daughter Maria – kept outside the door.

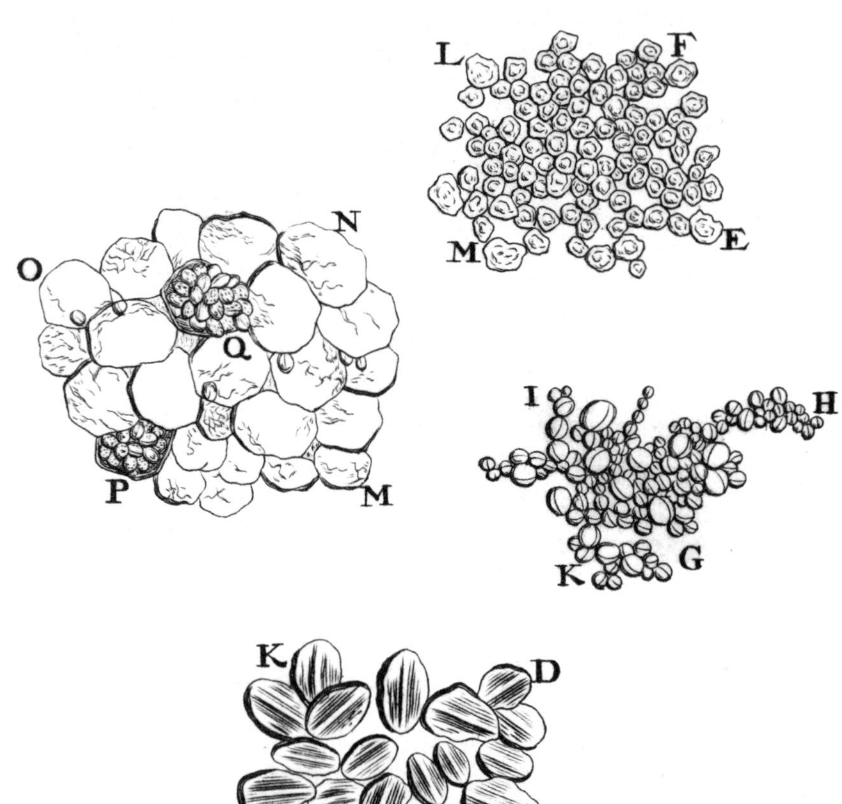
L
F
M
E
N
O
Q
P
M
I
H
K
G
K
D
L
C

NINE

Little Animals and Cells

In the autumn of 1723, Maria van Leeuwenhoek sent a small black, gilded cabinet to the Royal Society. The cabinet had five drawers, filled with little boxes that her father had lined with leather. In the boxes, two by two, were 26 silver microscopes, all made by Antoni.[1] In front of each lens he had affixed a specimen, as he always did, and for the sake of clarity he had included a register of all the items for the Society's fellows to admire. The specimens included red blood globules, a cochineal louse embryo, part of the crystalline humour of a whale's eye, the instrument used by a spider to spin thread, an ultra-thin membrane that covered a very small muscle and a double thread spun by a silk worm.[2]

The journey from Delft to London was not good for the small microscopes and, by the time the cabinet arrived at the Society, several of Antoni's specimens had broken loose. But it remained a splendid gift.[3] Antoni had prepared the whole cabinet with great care. It was a farewell gift, for after his death.

In the previous winter, Antoni – now ninety years old – had problems with strange palpitations in his midriff. The sensations were quite troublesome, but that did not stop him from clearly analysing and describing his own symptoms. Consequently, much later, the complaint was referred to as 'Van Leeuwenhoek's disease'.[4]

39 Antoni developed a breathing apparatus, with an opening at the top where air could enter and a mouthpiece at the bottom where the air could exit. He filled the middle section with a fluid that he hoped had healing powers, and through which the air passed before he breathed it in.

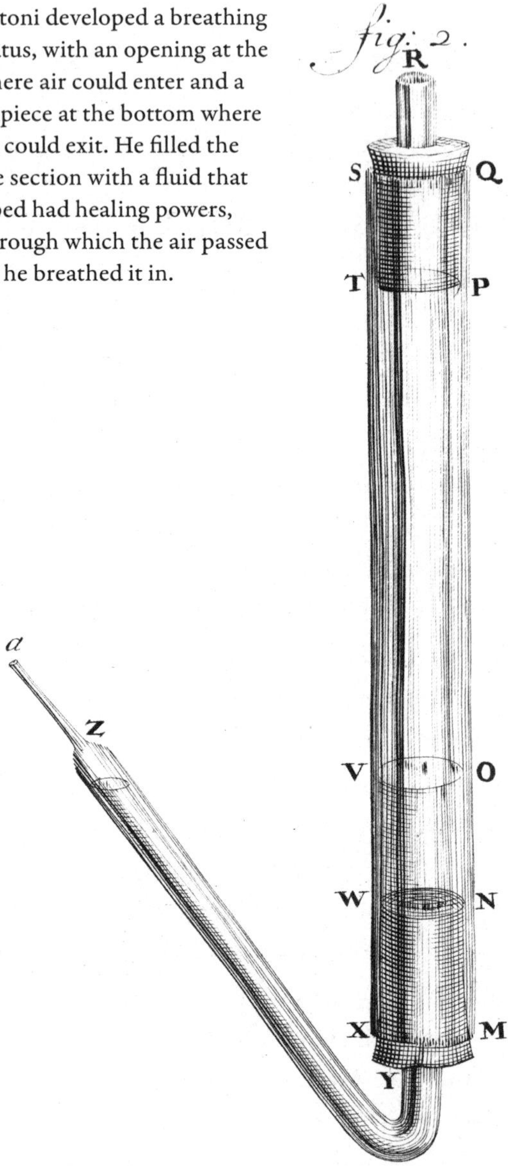

Because of his symptoms, he decided to investigate what a midriff looked like. So he had two brought to his house, the first from a year-old sheep and then one from a cow, and examined small sections through his lenses. It proved to be a thin muscle, the structure of which astounded Antoni, even after half a century of research. He saw such an unbelievable number of fibres, veins and particles, all bundled together in a minuscule fragment of matter, that it was beyond his comprehension.[5]

The unpleasant sensation in his midriff had disappeared after a while, but in June 1723 it returned. In the hope of putting a stop to it, he experimented with a mixture of nutmeg, mace, cloves and saffron, which he had steeped in alcohol for some time.[6] He allowed air to pass through the solution and then breathed it in, most likely with a breathing apparatus that he had earlier developed himself.[7] The apparatus consisted of a glass tube containing the alcohol solution and closed off at both ends with a cork. He stuck a thinner tube through each cork (see illus. 39), the upper one to allow air to enter and the lower one to serve as a mouthpiece. The air that entered at the top thus passed through the solution and the mouthpiece, and ended up in Antoni's lungs.[8]

And so he inhaled the air from the apparatus, not a little, but with great force, to stop the palpitations.[9] Once, it seemed to have worked and he felt better for a short time. But subsequent attempts had little effect.[10] And there was nothing else to be done to combat the disease. On 26 August 1723, two months before his 91st birthday, he died. His second wife Cornelia had died much earlier, in 1694. So Maria, herself now 66, remained behind alone in Het Gouden Hoofd.

Standardization

It was a great loss for the Royal Society, secretary James Jurin wrote to Maria; the Society had lost its most valuable correspondent.[11] But the cabinet with Antoni's microscopes was very much appreciated, as it gave the Society's fellows the opportunity to replicate his observations and see how good his lenses were.[12]

The Society, however, did very little with the gift. It had all the research and observation tips that Antoni had written about over the course of many years, a collection of loose specimens that he had sent, and now it also had 26 microscopes, with attached specimens.[13] Curious fellows thus had ample opportunity to continue his work. But no one made use of that opportunity.

That would partly be due to the difficulty of replicating Antoni's way of working, and of preparing the specimens in particular, which required a high degree of dexterity. Perhaps more importantly, Antoni's methods were difficult to standardize. And that was one of the main requirements of natural science, which was in those years in the process of taking shape: ideally researchers should be able to repeat exactly the same observations, wherever they were. They could then compare their results systematically and draw robust conclusions.

But with Antoni's work, that was easier said than done. His microscopes were difficult to transport, as shown by the fate of the specimens that came loose in the carefully designed cabinet. That was clearly not the best way to share observations. Another option, making a large number of identical microscopes of the type Antoni used available for a whole group of researchers, was also not feasible. It was difficult to produce multiple lenses of the same strength and quality by holding a glass thread in a flame. Moreover, there was as yet no standard method for establishing the comparative strength of lenses. So the question remained to what extent observations could be compared.

Antoni's death also marked the demise of his single-lens microscope. Multiple-lens instruments, including the type that Hartsoeker had already described in 1694, gained in popularity.[14] They comprised a tube with lenses and a screw thread, which the user could use to alter the distance between the lenses and the specimen. That made the microscopes suitable for different users and different objects.

Furthermore, for a time, scientific attention shifted back towards larger objects, of the kind that researchers like Hooke had examined before Antoni began to share his observations. In the years after his death, plant seeds and insects were particularly popular, but not the tiniest components that Antoni had dissected and examined.[15]

There was another development that Antoni the land surveyor would probably have appreciated more, if he had known about it: a standard measure for use by microscopists. This method was less variable than the grains of sand, hair and, later, grains of barley that he had used for many decades to measure little animals and globules. In the second half of the eighteenth century, microscopists began to use micrometres, a further step towards systematization and comparative research.[16]

Without males

In the meantime, a number of major questions that Antoni had addressed during his lifetime remained unanswered. In the first place, there was the matter of little seed animals and eggs. Antoni had died believing that the seed animals contained the necessary components to make young creatures, but in that respect he had always been an exception. There was simply too much evidence that the eggs also played a part. But how it all worked was still a mystery.

Moreover, there was a complicating factor that Antoni had seen with his own eyes in 1695 when inspecting his garden – which was incidentally not outside his house but most likely elsewhere within the Delft city walls.[17] There were cherry trees and redcurrant bushes, most of which were infested with lice. But strangely enough, he saw no evidence of eggs or male lice, so it was a puzzle to him how the insects reproduced.[18] Then he cut open a number of females and saw that they were carrying young lice in their bodies – young that had been created without the intervention of males.[19]

That was the start of a new, small-scale line of research, which led Antoni to conclude that some animals reproduced without mating – even though he himself had earlier declared that impossible.[20] We now call that phenomenon 'parthenogenesis', and it indeed occurs with the species of lice found in Antoni's garden.[21] To make the whole story more complicated, he also believed that he had seen parthenogenesis in the case of a substance called cochineal that had been imported from South America and which could produce red dye. Europeans had long been trying to determine what exactly cochineal was. Some suspected that it came from small animals, a hypothesis that Antoni had more or less confirmed with the help of his microscopes: it comprised the bodily components of insects.[22] And, he concluded in 1704, they were all females, often with young in their bodies. That led him to believe that, like the lice in his garden, they could reproduce without males.[23]

But in this he was mistaken. Male cochineal lice are small and fragile and live only for a short time, and the red dye is made from dried females – explaining why there were no males in Antoni's cochineal. Yet the males do play an essential role in reproduction.[24]

Through lack of sufficient research materials, he thus attributed an existing phenomenon – parthenogenesis – to the wrong species.

*

So there was confusion all round – with hindsight, but also in the eyes of his contemporaries. By no means everyone was convinced that parthenogenesis was possible.[25] And there was also no consensus on that other major issue, of spontaneous generation. Despite all of Antoni's evidence to the contrary, many scholars still believed that Aristotle had been right and that life could be created from nothing, or from rotting matter.[26] And that applied in particular to Antoni's minuscule animals.[27]

Just how mistaken this theory was started to become clear more than forty years after Antoni's death, when microscopists shifted their attention back to the smallest creatures. In 1765, Swiss scientist Horace-Bénédict de Saussure (1740–1799) observed how some of the smallest animals could divide to make two new living beings.[28] That would later prove to be a groundbreaking discovery, as this form of reproduction is typical for the minuscule creatures we know call single-cell organisms. It was now clear that even these minutest of organisms thus came from somewhere and did not generate spontaneously. That eventually led science to reject the theory of spontaneous generation, which Antoni had tried to prove false during his lifetime.

Around the time that Saussure saw the little animals dividing, Italian biologist Lazzaro Spallanzani (1729–1799) conducted an important experiment with that other form of reproduction, for which males and females were required. As test animals, he chose frogs, with which the eggs and the sperm come into contact outside the female's body. He dressed the male with miniature pants, which did not allow the sperm to pass through. Thus attired, the male frogs proved unable to generate new life, even though females and eggs were close by. But when Spallanzani spread sperm on the eggs, tadpoles developed. This led him to conclude that contact between egg and sperm was crucial.[29] And that settled the question: as a result of this kind of research, science took its leave of ovists, with their one-sided focus on eggs, and of spermists like Antoni. But the little seed animals he had researched so extensively continued to be of great importance in biology.

No heads or tails

A more fundamental step was a new insight that took shape mainly in the nineteenth century, but which had its roots in experiments like that conducted by Saussure. It was the realization that Antoni's little animals were completely different than he had thought. They were not miniature versions of the animals he saw around him every day, with heads, tails, legs and blood vessels, but creatures with an entirely different structure of their own: that of a cell.

The term 'cell' was already in use in microscopic research before the world became acquainted with Antoni's observations, as Hooke had used it in his *Micrographia* to describe the hollow spaces in cork.[30] But it acquired its present-day meaning – that of a building block of life – only much later. And experiments like Saussure's played an important role in that development, because they made it clear that the little animals behaved in a fundamentally different way than their larger counterparts, for example in how they reproduced.

In addition, in the eighteenth and nineteenth centuries, much work was done to produce microscopes that magnified clearly while distorting the image much less than especially the multi-lens microscopes of the seventeenth century had done. In the first few decades of the nineteenth century, many better and relatively easy-to-use microscopes came on the market, leading to giant leaps in the knowledge of cells.[31]

During the same period, researchers also came to understand the role of Antoni's red blood globules. They proved to transport oxygen, a crucial substance that became the focus of scientific attention in the late 1770s.

Little worms

Antoni would have been surprised at the discoveries of his successors. And he would have been astounded to learn that his little animals came to play a prominent role in medicine. That occurred in the second half of the nineteenth century, when it became clear that 'bacteria' and small parasites could cause all kinds of medical disorders.[32]

That understanding had its harbingers in earlier times. During Antoni's lifetime, Athanasius Kircher – the proponent of spontaneous generation and the ubiquitous seeds – had established a link between little animals and disease. In 1646, Kircher – who also worked with microscopes – had seen 'little worms' in the blood of fever patients.[33] He claimed in 1658 that they came from the rotting process and the seeds accompanying it, and that they were the cause of infectious diseases like the plague.[34] Exactly what kind of 'worms' Kircher saw is unclear, as his microscope did not magnify sufficiently to distinguish bacteria. But perhaps he saw single-cell organisms several decades before Antoni.[35]

It also seems very probable that Antoni knew of Kircher's work, as he wrote in 1679 that living creatures had been seen in blood in

Rome. As Kircher lived in Rome, it is possible that Antoni was talking about the little worms.[36] Antoni wanted to see them too and examined blood from his own hand, but without success. To be certain, he also examined blood from Cornelia, Maria and a maid-servant; in the last case, he was kind enough to wait until she was undergoing bloodletting, so that he would not have to cause her pain.[37] In these samples, too, he observed no little worms, but after some time he saw that the red globules began to clot, while still moving very slowly. They looked very much like living creatures, Antoni wrote, which could explain Kircher's 'discovery'.[38]

Kircher's disease-carrying little worms therefore did not exist, according to Antoni. Nor did he believe that little animals in general were the cause of illness. He had a completely different opinion on where diseases came from. He explained it a year before his death, on the basis of a new phenomenon that had come from Asia.[39] There, the *Philosophical Transactions* reported in 1714, a reliable method had been developed of preventing children from contracting the dangerous disease smallpox.[40] It started with a healthy young boy – girls were not mentioned here – who had contracted smallpox and had developed the associated blisters on his skin. On the twelfth or thirteenth day of the illness, a few of the blisters were cut open and the fluid collected in a clean glass. Next, with the help of a needle or knife, a few wounds were made in the skin of a healthy person who had never had smallpox, just deep enough to draw blood. A little of the fluid from the pox blisters was mixed with the blood seeping from the wounds, which were then covered for a couple of hours. After a time, blisters appeared around the wounds, which was a good sign, as anyone receiving this treatment could never contract smallpox.[41]

That was hopeful news and in London experiments were soon conducted with this forerunner of present-day inoculation. And they were a success, as James Jurin wrote to Antoni in 1722. The question was, what mechanism was at work here? Were there perhaps little animals in the smallpox fluid? And could Antoni perhaps investigate the matter further?[42] Antoni was not enthusiastic about this request, as he believed that smallpox was caused by thick blood, a common denominator in all diseases that produced a fever. In the case of smallpox patients, the blood would clot in the smallest blood vessels and be forced to the exterior, forming pox blisters. This could also lead to the whole blood circulation system being blocked up, and in some cases to the death of the patient. In Antoni's view, little animals played no part at all.[43]

We now know that smallpox is caused by viruses, which present-day biologists do not consider living creatures. In that sense, Antoni was right to believe that smallpox had nothing to do with little animals. Furthermore, he did not have the equipment to discover the viruses, as they were far too small to be visible through his lenses. But Antoni's response is important for another reason: it showed that his opinion on infectious diseases was far different from our theories.

In addition, he had little faith in the principle of inoculation, because the treatment was limited to a few local wounds, while smallpox created problems throughout the whole body. He thought that it might work if small wounds were made everywhere on the patient's body and the pox fluid applied to all of them. But he felt that it would be better for him to remain silent on the whole matter, as he was not an expert in this area. It was more a subject for doctors, who had studied diseases and treatments.[44]

More than little animals

Antoni thus looked differently at diseases, and especially at his little animals, than researchers in the nineteenth century. Yet, he would later be acclaimed for his discoveries in microbiology. That was clearly expressed in 1875, when admirers organized a 'festival of science' to mark the second centenary of what they considered Antoni's most important discovery: that of microscopic creatures.[45] That they were a year too late, as Antoni had already written about the little animals in the Berkelse Meer in 1674, did nothing to dampen the celebration.

In that jubilee year, moreover, the first biography of Antoni was published, written by his distant relative Pieter Jacob Haaxman, who had played with one of his microscopes as a child.[46] This was followed in 1932 by a new book, by Clifford Dobell, with a long and technical title that Antoni himself would not have understood, even if it had been in Dutch: *Antony van Leeuwenhoek and His 'Little Animals': Being Some Account of the Father of Protozoology and Bacteriology and His Multifarious Discoveries in These Disciplines*.[47]

Later biographies devoted more space to other aspects of Antoni's research, but in these publications too, cells received the greatest attention.[48] That also applies to this book. The red globules, the little seed animals and the little animals in pepper water have received more attention than the salt crystals, plant seeds and the structure of different kinds of wood, while Antoni bequeathed impressive observations on, for example, the 'pipes' in wood and their function to later biologists.[49] That is partly because of Antoni's surprise and delight in observing the little animals and globules, and partly because we now know that these discoveries would lead

40 Maria honoured her father with a tomb in the Oude Kerk in Delft.

to far-reaching changes in our worldview, and that cells would take centre stage in our thinking on the nature of life.

But this book aims to describe Antoni's own world, and not our judgements from three centuries later. That's why it is important to give Antoni himself the final word on what was important in his research – in the first place, through his collected letters, since they contain the texts that he found worth printing. And they reveal no clear preference, as they cover a rich variety of subjects. Take the *Send-Brieven*, his last, extensive collection of letters, in which he spoke of muscle fibres, the hairs of mice, bears and moles, plant seeds, a whale's eye, little animals attached to duckweed, sperm ducts, polychrest salt, the coconut and much more.[50]

The microscopes that Antoni sent to London contained a similar mix of objects, including a piece of the whale's eye referred to in the *Send-Brieven*, as did the hundreds of specimens auctioned after Maria's death. The collection did not include, however, the microscopes with the specimens that historians would later be very enthusiastic about: there were no little water animals or seed animals. That was probably due to the limited lifespan of many of the tiny creatures, and to transport problems: Antoni did not have the means to send the little animals to London alive, or to keep them until the auction that took place almost a quarter of a century after his death.

All of the areas of study that Antoni had covered over his fifty years of research were thus of equal importance to him. For him, it was not about a handful of breakthroughs, but the unexpected world of the minuscule that he had discovered and that he showed to others, through his words, drawings and demonstrations. To his great delight, that had brought him the recognition of scientists who, in his eyes, were worthy of great respect, such as the fellows of the Royal Society.

Society vice president Martin Folkes explained after Antoni's death what made his observations so exceptional. It was the care he devoted to his lenses, Folkes wrote in *Philosophical Transactions*, which made them 'exceedingly clear' and gave them the correct 'power of magnifying' for his objects.[51] But no less important were his power of judgement and the devotion and skill with which he had prepared his specimens and repeatedly examined them for many years.[52] That put him leagues ahead of all other researchers, and that is why they had to take his work very seriously – even if they were unable to replicate his observations. They had to do their utmost if they wanted to achieve the same high level of perfection as Antoni, something his fellow members of the Royal Society failed to achieve in practice.[53]

Maria, too, felt that her father deserved recognition for his achievements. In 1739, she had a monument placed in the Oude Kerk in Delft, where he had been buried sixteen years earlier.[54] It bore his portrait and the following text – in Latin, which Antoni had never mastered and which made him always something of an outsider. By honouring her father in the language of science, Maria was making it clear to all learned scholars how much her father was worthy of their praise:

> To the honourable and eternal memory of Anthony van Leeuwenhoek, Fellow of the Royal Society in England.
>
> Who revealed the mysteries of nature and the secrets of the natural sciences with the aid of microscopes, designed and constructed with remarkable skill by his own hand, through persistent industry and research, and described them in the Dutch language, such that he was considered exceptionally meritorious around the whole world.[55]

REFERENCES

Introduction: Lost Wonders

1 *Catalogus van het vermaarde cabinet van vergrootglasen, met zeer veel moeite, en kosten in veele jaren geïnventeert, gemaakt, en nagelaten door wylen den heer Anthony van Leeuwenhoek, in zyn Ed: Leeven Lid van de Koninglyke Societeit der Wetenschappen in Londen* (Delft, 1747).

2 J. G. Kerckherdere and Arnold Hoogvliet (translator), 'Lofdicht op den zeer vermaarden en in kunst en wetenschap ervaren Antoni van Leeuwenhoek', in *Lauwerkranssen, gevlochten voor den heer Antoni van Leeuwenhoek, groot wysgeer, lidt der Koningklyke Gemeenschap te Londen* (Rotterdam, 1717), 12, 13.

3 Huib J. Zuidervaart and Douglas Anderson, 'Antony van Leeuwenhoek's microscopes and other scientific instruments: New information from the Delft archives', *Annals of Science* 73, 3 (2 July 2016), 257–88, in particular 270; 'Bekentmaking', *Hollandsche Historische Courant*, 16 May 1747, 2.

4 Zuidervaart and Anderson, 'Antony van Leeuwenhoek's microscopes and other scientific instruments', 281–3.

5 Ibid., 281; Jan de Vries and A. M. van der Woude, *Nederland 1500–1815: De eerste ronde van moderne economische groei*, 3rd edn (Amsterdam, 2005), 653.

6 Zuidervaart and Anderson, 'Antony van Leeuwenhoek's microscopes and other scientific instruments', 281; *Catalogus van het vermaarde cabinet van vergrootglasen*, 19, 21, 27, 29.

7 Zuidervaart and Anderson, 'Antoni van Leeuwenhoek's microscopes and other scientific instruments', 275.

8 Ibid., 277.

9 P. J. Haaxman, *Antoni van Leeuwenhoek. De ontdekker der infusoriën 1675–1875* (Leiden, 1875), 33, 34.

10 Four of the microscopes are in Rijksmuseum Boerhaave in Leiden, one is in Utrecht University Museum, one in Planetarium Zuylenburgh in Oud-Zuilen and two in the Deutsches Museum in Munich. One is still in the possession of the Haaxman family, and one was sold to an unknown buyer in 2009. One possibly genuine microscope is in the

possession of the Zoological Museum in Antwerp, now on loan to Ghent University Museum; Tiemen Cocquyt, 'De identificatie van een zilveren microscoopje van Antoni van Leeuwenhoek (1632–1723)', *Studium* 8, 4 (24 May 2016), 198; GUM, 'Deelcollectie Geschiedenis van de Wetenschappen', www.gum.gent, accessed 5 December 2022.

11 Haaxman, *Antoni van Leeuwenhoek. De ontdekker der infusoriën*.

1: Child of Honest Parents

1 Zacharias Conrad von Uffenbach, *Merkwürdige Reisen durch Niedersachsen, Holland und Engelland*, vol. III (Ulm, 1754), 335, 349.
2 Joris Geesteranus, 'Inventaris nalatenschap Maria van Leeuwenhoek', 26 July 1745, Old Notarial Archives Delft, 1574–1842, Delft Municipal Archives, 161-2791 [118, 119].
3 Ibid.
4 Von Uffenbach, *Merkwürdige Reisen durch Niedersachsen, Holland und Engelland*, vol. III, 349–50.
5 Ibid., 355.
6 Ibid., 357.
7 Ibid., 355.
8 Ibid., 350.
9 Ibid., 355, 356.
10 Ibid., 357–8.
11 Ibid., 352–3.
12 Ibid., 360.
13 Ibid., 350, 357, 358, 360.
14 Reinier Boitet, *Beschryving der stadt Delft, behelzende een zeer naaukeurige en uitvoerige verhandeling van deszelfs eerste oorsprong, benaming, bevolking, aanwas, gelegenheid, prachtige en kunstige gedenkstukken en zeltzaamheden. Nevens derzelver voorregten, handvesten, previlegien, en regeeringsvorm: alles 't zamengestelt en getrokken uit oude handtschriften, memorien, en brieven, en met zeer veele echte bewysstukken (te vooren noit gedrukt) bevestigt: door verscheide liefhebbers en kenners der Nederlandsche oudheden* (Delft, 1729), 765.
15 W. H. van Seters, 'Leeuwenhoek's afkomst en jeugd', *Biologisch Jaarboek* 19 (1952), 123–84, in particular 125–51.
16 These were Grietge's paternal grandfather, Sebastiaen van den Berch, and his father-in-law and Grietge's great-grandfather, Jan Jacobsz in het Ossenhooft. Ibid., 129, 131.
17 Ibid., 128, 142; Antoni van Leeuwenhoek, letter no. 325 [XXII], 16 May 1716, *Alle de brieven* 17 (London, 2018).
18 Thera Wijsenbeek-Olthuis, *Achter de gevels van Delft: Bezit en bestaan van rijk en arm in een periode van achteruitgang (1700–1800)* (Hilversum, 1987), 57.
19 Van Seters, 'Leeuwenhoek's afkomst en jeugd', 142.

20 Ibid., 143.
21 Jacob's parents Sebastiaen Cornelisz and Neeltje Jansdr married on 23 January 1556. Jacob himself married on 24 January 1588 and was admitted to the brewers' guild in 1589. It can therefore be assumed that Jacob was born between 1556 and 1568.
22 Van Seters, 'Leeuwenhoek's afkomst en jeugd', 143.
23 'Ondertrouwboeken Nieuwe Kerk Delft 1587–1626, deel 6. April 1621–27 december 1626', Delft Municipal Archives, 14-6, 18v.
24 Douglas Anderson, 'Thonis Philips Leeuwenhoek', *Lens on Leeuwenhoek*, https://lensonleeuwenhoek.net/content/thonis-philips-leeuwenhoek, accessed 13 December 2022; 'Legger van de verponding op huizen en molens te Delft, voor ontvanger Adriaen Claesz Moeijt (1620 aangelegd, met aantekeningen tot 1634)', Delft Municipal Archives, 1-4013; 'Kohier van de verponding op huizen en erven binnen de stad volgens het redres-generaal van 1632 (met aantekeningen tot 1656)', Delft Municipal Archives, 1-4017; 'Testament Thonis Philipsz Leeuwenhoek sr. bij notaris Guillaume de Graeff', 2 April 1643, Old Notarial Archives, 161–171 f 32 (55v), Delft Municipal Archives.
25 Gerrit Verhoeven, *De derde stad van Holland: Geschiedenis van Delft tot 1795* (Zwolle, 2015), 357–61.
26 'Doopboeken Nieuwe Kerk 9 oktober 1616–31 maart 1624', Delft Municipal Archives, 14-45, 128v; 'Doopboeken Oude Kerk, 1 augustus 1624–31 juli 1642', Delft Municipal Archives, 14-8, 19v; 'Doopboeken Nieuwe Kerk april 1624–10 augustus 1636', Delft Municipal Archives, 14-55, 57, 84.
27 Tomb of Antoni van Leeuwenhoek, Oude Kerk, Delft; 'Doopboeken Nieuwe Kerk, april 1624–10 augustus 1636', Delft Municipal Archives, 14-15, 169v; 'Doopboeken Nieuwe Kerk 14 augustus 1636–1649', Delft Municipal Archives, 14-56, 11v.
28 'Doopboeken Nieuwe Kerk april 1624–10 augustus 1636'.
29 Barbara was baptized on 20 December 1629 in the Nieuwe Kerk: 'Doopboeken Nieuwe Kerk april 1624–10 augustus 1636'.
30 Lesley Robertson et al., *Van Leeuwenhoek: groots in het kleine* (Amsterdam, 2014), 3.
31 Letter no. 26 [18], 9 October 1676, *Alle de brieven* 2, 84.
32 Anderson, 'Thonis Philips Leeuwenhoek'.
33 Letter no. 106 [61], 25 May 1688, *Alle de brieven* 7, 146.
34 This warehouse was on the west side of the Oude Delft. In 1653 a second warehouse was built on the other side of the canal. W. F. Weeve, 'Het Oostindisch pakhuis te Delft', in *Delft en de Oostindische Compagnie*, ed. H. L. Houtzager (Amsterdam, 1987), 131–50, in particular 141.
35 Verhoeven, *De derde stad van Holland*, 296.
36 Huib J. Zuidervaart and Marlise Rijks, '"Most rare workmen": Optical practitioners in early seventeenth-century Delft', *British Journal for the History of Science* 48, 1 (March 2015), 76.

37 Ibid., 61.
38 A. Th. van Deursen, 'Mensen van klein vermogen. Het kopergeld van de Gouden Eeuw', in *De Gouden Eeuw compleet* (Amsterdam, 2010), 142.
39 E. P. de Booy, 'Het onderwijs in Delft van 1572 tot het midden van de 17e eeuw', in *De stad Delft. Cultuur en maatschappij van 1572 tot 1667*, ed. I. Spaander (Delft, 1981), 112–20.
40 E. P. de Booy, *Kweekhoven der wijsheid. Basis- en vervolgonderwijs in de steden van de provincie Utrecht van 1580 tot het begin der 19e eeuw* (Zutphen, 1980), 26.
41 De Booy, 'Het onderwijs in Delft van 1572 tot het midden van de 17e eeuw', 118.
42 De Booy, *Kweekhoven der wijsheid*, 46.
43 Carel de Gelliers, *Trap der jeugd, ofte Perfecte maniere om de Jonge kinderen en Oude Personen met Fondament te leeren lezen en schrijven. Daar beneffens een kort onderrigt om de penne wel te snijden, te houden en te voeren. Vercierd met gebeden, dankzeggingen en Christelijke spreuken* (1640), cited in Annemarie Toorn and Marijke Spies, in cooperation with Sietske Hoogerhuis, 'Christen Jeugd, leerd Konst en Deugd. De zeventiende eeuw', in *De hele Bibelebontse berg. De geschiedenis van het kinderboek in Nederland & Vlaanderen van de middeleeuwen tot heden*, ed. Harry Bekkering et al. (Amsterdam, 1989), 105–68, in particular 115.
44 The first – anonymous – version of the book was published in 1614 under the title *Spiegel der jeught*. In 1620 a new version was published with a subtitle: *oft Spaensche tyrannye, gheschiet in Nederlandt, waer in te sien is de onmenschelijcke ende wreede handelingen der Spangiaerden*. Toorn and Spies, 'Christen Jeugd, leerd Konst en Deugd. De zeventiende eeuw', 127–9.
45 De Booy, 'Het onderwijs in Delft van 1572 tot het midden van de 17e eeuw', 117–18.
46 Ibid., 118.
47 'Testament Thonis Philipsz Leeuwenhoek sr. bij notaris Guillaume de Graeff'.
48 De Booy, *Kweekhoven der wijsheid*, 48. Antoni van Leeuwenhoek letter no. 4, 5 April 1674, *Alle de brieven* 1, 66; letter no. 2, 15 August 1673, *Alle de brieven* 1, 60.
49 De Booy, *Kweekhoven der wijsheid*, 52.
50 Willem Bartjens, Danny Beckers and Marjolein Kool, *De cijfferinghe (1604). Het rekenboek van de beroemde schoolmeester Willem Bartjens. Ingeleid door Danny Beckers en Marjolein Kool* (Hilversum, 2004), 133.
51 Danny Beckers and Marjolein Kool, 'Hoofdstuk 5. De rekenboeken van Willem Bartjens', in Bartjens, Beckers and Kool, *De cijfferinghe*, 54–76, in particular 63.
52 William P. Berlinghoff, Fernando Q. Gouvêa and Desiree van den Bogaart, *Wortels van de wiskunde: Een historisch overzicht voor leraren en anderen* (Amsterdam, 2016), 94–9.

53 Nelleke Bakker, Jan Noordman and Marjoke Rietveld-Van Wingerden, *Vijf eeuwen opvoeden in Nederland: idee en praktijk 1500–2000* (Assen, 2010), 124.

54 On 27 February 1639 a Jacob Phillipsz was buried in the Nieuwe Kerk in Delft. 'Begraafboeken Oude en Nieuwe Kerk, mei 1628–1 januari 1644', Delft Municipal Archives, 14-00038. E. W. van den Burg and G. J. Leeuwenhoek suggest in '(Van) Leeuwenhoek', *Kronieken* (1995), that this was Thonis's younger brother. That is not feasible, as the buried individual lived in the Molenstraat, and there is no evidence that the family ever lived there. The deceased was probably Jacob Philips, a carpenter who was married on 23 September 1628 and was at that time already living in the Molenstraat: 'Ondertrouwboeken Nieuwe Kerk 1626–1811', Delft Municipal Archives, 14-68, 54v.

Both Jacob and Maria must have died before grandfather Thonis drew up his will, in 1643. In the will, he left money for four children from the family. That must have been Margrieta, Geertruyt, Neeltje and Thonis, as they were still alive: 'Testament Thonis Philipsz Leeuwenhoek sr. bij notaris Guillaume de Graeff'.

55 'Begraafboeken Oude en Nieuwe Kerk, 4 januari 1644–30 januari 1656', Delft Municipal Archives, 14-39, 207v.

56

Year	Deaths according to the funeral books
1630	179
1631	188
1632	169
1633	196
1634	183
1635	586
1636	473
1637	192
1638	216
1639	191
1640	214

Source: Delft funeral books at https://wiewaswie.nl, accessed 9 March 2021.

In the plague years 1635–6 there were three hundred excess deaths, or 1.3 per cent of the population. That is in accordance with the estimate that in a plague year the disease led to a death rate of 0.35 per cent to a little over 3 per cent. F.W.A. van Poppel, 'Het komen en gaan van ziekten', in H.F.P. Hillen, E. S. Houwaart and F. G. Huisman, eds, *Medische geschiedenis: Ziekte, kennis, dokter en patiënt, gezondheidszorg en maatschappij* (Houten, 2018), 3–18, in particular 7.

57 Leo Noordegraaf and Gerrit Valk, *De gave Gods: de pest in Holland vanaf de late middeleeuwen*, 2nd revd edn (Amsterdam, 1996), 31.

58 Peter Elmer, ed., *The Healing Arts: Health, Disease and Society in Europe, 1500–1800* (Manchester, 2004), 63–4.

59 Mary Lindemann, *Medicine and Society in Early Modern Europe* (Cambridge, 2020), 18.
60 Hannah Newton, *The Sick Child in Early Modern England, 1580–1720* (Oxford, 2012), 40.
61 Lindemann, *Medicine and Society in Early Modern Europe*, 13.
62 Newton, *The Sick Child in Early Modern England*, 39.
63 Elmer, *The Healing Arts*, 63.
64 Ibid.
65 Lindemann, *Medicine and Society in Early Modern Europe*, 218.
66 Ibid., 53, 218.
67 'Begraafboeken Oude en Nieuwe Kerk, mei 1628–1 januari 1644', Delft Municipal Archives, 14-38, 92v.
68 Douglas Anderson, 'Jan Jacobs de Molijn', *Lens on Leeuwenhoek*, https://lensonleeuwenhoek.net, accessed 14 December 2022; 'Ondertrouwboek Gerecht, 1637–1645', Delft Municipal Archives, 14-1225, 67.

When De Molijn announced the banns for his second marriage on 13 July 1624, he was registered as a bailiff. That probably meant that he collected taxes for the city. 'Ondertrouwboek Gerecht, 10 maart 1618–26 september 1626', Delft Municipal Archives, 14-124, 130 v.

In addition, Jacop Molijn was registered in the 'Meesterboeck' of the St Lucas Guild for the years 1613–49 as a painter. D. O. Obreen, 'Het Sint Lucas-Gild te Delft', *Archief voor Nederlandsche kunstgeschiedenis*, vol. 1 (Rotterdam, 1877), 1–120, in particular 4.
69 Douglas Anderson, 'Stepfather Jacob Jans de Molijn appeared before Weeskamer to appoint children's guardians', *Lens on Leeuwenhoek*, https://lensonleeuwenhoek.net, accessed 14 December 2022; Archief Weeskamer Delft, Delft Municipal Archives, 454 f. 373; 'Ondertrouwboeken Nieuwe Kerk 1626–1811', Delft Municipal Archives, 14-69, 127v.
70 Boitet, *Beschryving der stadt Delft*, 765.
71 W. H. van Seters, 'Can Antoni van Leeuwenhoek have attended school at Warmond?', in *Antoni van Leeuwenhoek 1632–1723: Studies on the Life and Work of the Delft Scientist Commemorating the 350th Anniversary of His Birthday*, ed. L. C. Palm and H.A.M. Snelders (Amsterdam, 1982), 7–8.
72 Ibid., 5.
73 Bakker, Noordman and Rietveld-Van Wingerden, *Vijf eeuwen opvoeden in Nederland*, 530–31.
74 Ibid., 531.
75 W. H. van Seters, 'Antoni van Leeuwenhoek in Amsterdam', *Notes and Records of the Royal Society of London* 9, 1 (31 October 1951), 36–45, in particular 36.
76 Boitet, *Beschryving der stadt Delft*, 765.
77 Van Seters, 'Antoni van Leeuwenhoek in Amsterdam', 38.
78 Ibid., 41, 44–5.

2: Meeting the Minuscule

1 J. Heniger, 'Antoni van Leeuwenhoek en zijn diploma van de Royal Society', *Tijdschrift voor de geschiedenis der geneeskunde, natuurwetenschappen, wiskunde en techniek*, 1 (1978), 159; Thomas Birch, *History of the Royal Society of London*, vol. III (London, 1757), 338, 346, 347, 349, 352.
2 'Ondertrouwboek 1650–1656', Delft Municipal Archives, 14-27, 64v.
3 Douglas Anderson, 'Inherited property on Oosteinde from first wife Barbara de Meij's family', *Lens on Leeuwenhoek*, https://lensonleeuwenhoek.net, accessed 14 December 2022.
4 A. Schierbeek, *Antoni van Leeuwenhoek. Zijn leven en zijn werken* (Lochem, 1950), 16–17.
5 Ibid., 16.
6 Douglas Anderson, 'Hippolytusbuurt 3, Leeuwenhoek's home and laboratory', *Lens on Leeuwenhoek*, https://lensonleeuwenhoek.net, accessed 14 December 2022.
7 Stedelijk Museum Het Prinsenhof, *De stad Delft. Cultuur en maatschappij van 1572 tot 1667*, ed. in I. Spaander (Delft, 1981), 16, 17.
8 Philips was baptized on 15 September 1655: 'Doopboeken Oude Kerk, 1 August 1642–27 August 1658', Delft Municipal Archives, 14-9, 173 v.
9 Philips was buried on 27 October 1655: 'Begraafboeken Oude en Nieuwe Kerk , 4 januari 1644–30 januari 1656', Delft Municipal Archives, 14-39, 124v.
10 The second son Philips was baptized on 25 January 1663: 'Doopboeken Nieuwe Kerk, 1661–1682', Delft Municipal Archives, 14-58 (25v). He was buried on 14 February 1663: 'Begraafboeken Oude en Nieuwe Kerk, 1 februari 1655–30 oktober 1666', Delft Municipal Archives, 14-40, 63.
11 The third Philips was baptized on 9 August 1664: 'Doopboeken Oude Kerk, 10 juni 1664–31 December 1684', Delft Municipal Archives, 14-11, 2v. He was buried on 27 June 1666: 'Begraafboeken Oude en Nieuwe Kerk, 1 februari 1655–30 oktober 1666', 107v.
12 Margrieta was baptized on 29 September 1658: 'Doopboeken Oude Kerk, 1 september 1658–18 juni 1664', Delft Municipal Archives, 14-10, 2v. She was buried on 7 October 1658: 'Begraafboeken Oude en Nieuwe Kerk, 1 februari 1656–30 oktober 1666', Delft Municipal Archives, 14-40, 20v.
13 Douglas Anderson, 'The early civic career of Anthony van Leeuwenhoek: The case of Sijmon Bourbon [1667–70]', article under preparation.
14 Schierbeek, *Antoni van Leeuwenhoek*, 18–19.
15 Antoni was a wealthy man when he died and his estate included considerable capital, in the form of real estate, annuities and bonds: Joris Geesteranus, 'Inventaris nalatenschap Maria van Leeuwenhoek', 26 July 1745, Old Notarial Archives Delft, 1574–1842, Delft Municipal Archives, 161-2791.
16 Douglas Anderson, 'Still going strong: Leeuwenhoek at eighty', *Antonie van Leeuwenhoek. Journal of Microbiology* 106 (2014), 9, 11.

17 As a general district supervisor, Antoni earned 50 guilders a year: Schierbeek, *Antoni van Leeuwenhoek*, 19.
18 Anderson, 'The early civic career of Anthony van Leeuwenhoek', 14.
19 Ibid., 11.
20 Ibid., 18.
21 Ibid., 9–10.
22 Ibid., 6.
23 Antoni was paid 42 guilders and 6 stuivers. Anderson, 'The early civic career of Anthony van Leeuwenhoek', 19.
24 Ibid., 20–22.
25 Ibid., 23.
26 'Doopboeken Nieuwe Kerk april 1624–10 augustus 1636', Delft Municipal Archives, 14-55, 119.
27 Gilles Aillaud et al., *Vermeer* (Paris, 1987), 57–61.
28 Klaas van Berkel, *Citaten uit het boek der natuur: opstellen over Nederlandse wetenschapsgeschiedenis* (Amsterdam, 1998), 116.
29 'Ondertrouwboeken Nieuwe Kerk 1626–1811', Delft Municipal Archives, 14-70 (117v). That Antoni was in contact with his stepbrother Jan Jacobs de Molijn can be seen from the fact that, together with his sister Maria, Jan stood as guarantor for the loan that Antoni took out to buy Het Gouden Hoofd: Douglas Anderson, 'Jan Jacobs de Molijn', *Lens on Leeuwenhoek*, https://lensonleeuwenhoek.net, accessed 14 December 2022; 'Minuutakten notaris Andries Bogaert 1654–1658', Delft Municipal Archives, 161-1888 (70v).
30 D. O. Obreen, 'Het Sint Lucas-Gild te Delft', *Archief voor Nederlandsche kunstgeschiedenis*, vol. 1 (Rotterdam, 1877), 1–120, in particular 6.
31 *The Geographer*, *c.* 1669, is in the possession of the Städel Museum, Frankfurt am Main, https://sammlung.staedelmuseum.de, accessed 12 August 2022. *The Astronomer*, *c.* 1668, is in the possession of the Louvre, https://collections.louvre.fr, accessed 12 August 2022. Van Berkel, *Citaten uit het boek der natuur*, 116–17.
32 Luuc Kooijmans, *Gevaarlijke kennis: Inzicht en angst in de dagen van Jan Swammerdam* (Amsterdam, 2007), 74–5 and 79–80.
33 G. A. Lindeboom, 'Reinier de Graaf (1641–1673)', *Nederlands Tijdschrift voor Geneeskunde* 118, 21 (1974), 791.
34 Ibid., 791.
35 Evan Ragland, 'Experimenting with chymical bodies: Reinier de Graaf's investigations of the pancreas', *Early Science and Medicine* 13, 6 (2008), 616.
36 Lindeboom, 'Reinier de Graaf (1641–1673)', 792.
37 Ragland, 'Experimenting with chymical bodies', 619.
38 Reinier de Graaf, *Tractatus anatomico-medicus de succi pancreatici natura et usu* (Leiden, 1671), 31.
39 Rolf Willach, 'The long road to the invention of the telescope', in *The Origins of the Telescope*, ed. Albert van Helden et al. (Amsterdam, 2010), 93–114, in particular 94–5.

40 George Sines and Yannis A. Sakellarakis, 'Lenses in antiquity', *American Journal of Archaeology* 91, 2 (1987), 191–6, in particular 193.
41 Willach, 'The long road to the invention of the telescope', 104–6.
42 Ibid., 107.
43 Charles Singer, 'The dawn of microscopical discovery', *Journal of the Royal Microscopical Society* (1915), 317–40, in particular 318.
44 Willach, 'The long road to the invention of the telescope', 111–12.
45 Tiemen Cocquyt, 'Positioning Van Leeuwenhoek's microscopes in 17th-century microscopic practice', *FEMS Microbiology Letters* 369, 1 (2022), 1.
46 Silvio A. Bedini, 'Galileo and scientific instrumentation', in *Reinterpreting Galieo*, Studies in Philosophy and the History of Philosophy 15, ed. William A. Wallace (Washington, DC, 1986), 127–53, in particular 147.
47 Paolo Galluzzi and Peter Mason, *The Lynx and the Telescope: The Parallel Worlds of Federico Cesi and Galileo* (Leiden and Boston, MA, 2017), 385.
48 Ibid., 386.
49 Eric Jorink, '"These wonderful glasses": Dutch humanists and the microscope, 1620–1670', in *Who Needs Scientific Instruments? Conference on Scientific Instruments and Their Users, 20–22 October 2005*, ed. Bart Grob and Hans Hooijmaijers (Leiden, 2006), 115.
50 Constantijn Huygens and Chris L. Heesakkers, *Mijn jeugd* (Amsterdam, 1987), 132.
51 Jorink, 'These wonderful glasses', 116; Huygens and Heesakkers, *Mijn jeugd*, 132.
52 J. de Wyck to Christiaan Huygens, 27 October 1654, letter 202 in Christiaan Huygens, *Oeuvres complètes*, vol. 1: *Correspondance, 1638–1656* (The Hague, 1888), 302; Huib J. Zuidervaart and Marlise Rijks, '"Most rare workmen": Optical practitioners in early seventeenth-century Delft', *British Journal for the History of Science* 48, 1 (2015), 69.
53 Zuidervaart and Rijks, 'Most rare workmen', 69.
54 'About nine or ten years since Dr. Graaf opened in my presence the vein of a Dog', wrote Antoni van Leeuwenhoek on 14 January 1678. That must therefore have been around 1668 or 1669. Letter no. 37 [23], 14 January 1678, *Alle de brieven* 2, 310.
55 Letter no. 37 [23], 310.
56 L. C. Palm, 'De Graafs invloed op Van Leeuwenhoek', in *Reinier de Graaf, 1641–1673: in sijn leven nauwkeurig ontleder en gelukkig geneesheer tot Delft; bundel opstellen, verschenen bij gelegenheid van de herdenking van de 350ste geboortedag van Reinier de Graaf*, ed. Hendrik Leonard Houtzager (Rotterdam, 1991), 41–52, in particular 45.
57 Letter no. 37 [23], 312.
58 Matthew Cobb, *Generation: The Seventeenth-Century Scientists Who Unraveled the Secrets of Sex, Life, and Growth*, 1st U.S. edn (New York, 2006), 171.

59 Regnerus de Graaf, *Alle de wercken, so in de ontleed-kunde, als andere deelen der medicyne* (Amsterdam, 1686), 320.

60 Grietge (written here as 'Grietje') van den Berch was buried in the Oude Kerk on 3 September 1664 and Barbara de Meij on 14 July 1666, in the same church. 'Begraafboeken Oude en Nieuwe Kerk, 1 februari 1655–30 oktober 1666', Delft Municipal Archives, 14-40, 75v, 108.

61 In 1664 he was still supplying goods, as can be seen from 'Inventaris van den boedel en goederen van Elisabeth Cornelisdr door notaris Johannes Ranck', Minute 1163, in 'Minuten van allerlei akten', Delft Municipal Archives, 161-2119 (scan from public website of Delft Municipal Archives, 174). From 1667 Antoni no longer paid the tax on goods that was compulsory for a vendor: Anderson, 'The early civic career of Anthony van Leeuwenhoek: The case of Sijmon Bourbon [1667–70]', 11.

62 Letter no. 11 [6], 7 September 1674, *Alle de brieven* 1, 158.

63 Letter no. 11 [6], 158.

64 Hans Bots, *De Republiek der Letteren: De Europese intellectuele wereld 1500–1760* (Nijmegen, 2018), 54–5.

65 When Barbara's parents announced the banns for their marriage on 5 June 1622, her father was recorded as coming from 'Norwits' (Norwich). 'Ondertrouwboek Gerecht, 10 maart 1618–26 september 1626', Delft Municipal Archives, 14-124, 90.

66 Robert Hooke, *Micrographia: or Some Physiological Descriptions of Minute Bodies Made By Magnifying Glasses With Observations and Enquiries thereupon* (London, 1665), preface.

67 Letter no. 103 [58], 9 September 1687, *Alle de brieven* 7, 64.

68 Tiemen Cocquyt, Marvin Bolt and Michael Korey, 'Hudde en zijn gesmolten microscooplensjes', *Studium* 11, 1 (10 October 2018), 78.

69 Hooke, *Micrographia*, preface.

70 Clifford Dobell, *Antony van Leeuwenhoek and His 'Little Animals', Being Some Account of the Father of Protozoology and Bacteriology and His Multifarious Discoveries in These Disciplines*, new edn (New York, 1960), 40–41.

71 Letter no. 1 [1], 28 April 1673 *Alle de brieven* 1, 32.

72 Letter no. 1 [1], 32.

73 Ad Leerintveld, 'Huygens en Van Leeuwenhoek', Brieven Constantijn Huygens online, https://brievenconstantijnhuygens.net, accessed 22 July 2022.

74 Letter no. 8 [4], 1 June 1674, *Alle de brieven* 1, 98.

75 Constantijn Huygens Jr to Christiaan Huygens, 13 August 1680, in Christiaan Huygens, *Oeuvres complètes*, vol. VIII: *1676–84* (The Hague, 1899), 295–6, in particular 296.

3: Globules and Stalks

1 Letter no. 26 [18], 9 October 1676, *Alle de brieven* 2, 130; Lodewijk C. Palm, 'Antoni van Leeuwenhoeks kristallographische und mineralogische Untersuchungen', *Montanmedizin und Bergbauwissenschaften: Hallesches Symposium* (1987), 115.
2 Evan Ragland, 'Experimenting with chymical bodies: Reinier de Graaf 's investigations of the pancreas', *Early Science and Medicine* 13, 6 (2008), 615–64, in particular 625.
3 Marian Fournier, *The Fabric of Life: The Rise and Decline of Seventeenth-Century Microscopy* (Enschede, 1991), 127.
4 Letter no. 4, 5 April 1674, *Alle de brieven* 1, 66–8.
5 Letter no. 50, 11 July 1679, *Alle de brieven* 3, 88, 90.
6 Ibid., 90, 92.
7 Ibid., 94.
8 Letter no. 52, 14 November 1679, *Alle de brieven* 3, 114.
9 Ibid.
10 Letter no. 107 [62], 6 July 1688, *Alle de brieven* 7, 258, 268; letter no. 149 [91], 20 July 1695, *Alle de brieven* 22; letter no. 83 [44], 23 January 1685, *Alle de brieven* 5, 70.
11 Letter no. 246, 1 February 1704, *Alle de brieven* 294, 296, 300.
12 Letter no. 253, 13 December 1704, *Alle de brieven* 15, 88.
13 Letter no. 253, 90, 92, 94.
14 Hugh Aldersey-Williams, *Dutch Light: Christiaan Huygens and the Making of Science in Europe* (London, 2020), 370, 377.
15 H.A.M. Snelders, 'Christiaan Huygens and Newton's theory of gravitation', *Notes and Records of the Royal Society of London* 43, 2 (1989), 212.
16 Letter no. 168 [101] 10 July 1696, *Alle de brieven* 11, 294.
17 H.A.M. Snelders, 'Antoni van Leeuwenhoek's mechanistic view of the world', in *Antoni van Leeuwenhoek 1632–1723: Studies on the Life and Work of the Delft Scientist Commemorating the 350th Anniversary of His Birthday*, ed. L. C. Palm and H.A.M. Snelders (Amsterdam, 1982), 57–78, in particular 73.
18 Letter no. 2, 15 August 1673, *Alle de brieven* 1, 56.
19 Ibid.
20 Ibid.
21 Ibid.
22 Ibid.
23 Robert Boyle, 'To the Lord of Dungarvan, my honoured and dear nephew', in *The Works of Robert Boyle. Electronic Edition*, vol. 1: *Publications to 1660*, ed. Michael Hunter and Edward B. Davis (London, 1999), 157–300, in particular 165.
24 Steven Shapin and Simon Schaffer, *Leviathan and the Air-Pump: Hobbes, Boyle, and the Experimental Life* (Princeton, NJ, 1989), 44–5.

25 Hooke, *Micrographia*, preface.
26 Letter no. 19 [13], 20 December 1675, *Alle de brieven* 1, 336.
27 Robert Boyle, 'Shewing the occasion of making this new essay – instrument, together with the hydrostatical principle 'tis founded on', *Philosophical Transactions* 10, 115 (1675), 330, 340.
28 Letter no. 19 [13], 336.
29 Ibid., 336.
30 Boyle, 'Shewing the occasion', 329–30.
31 Hooke, *Micrographia*, 125.
32 Letter no. 1 [1], 28 April 1673, *Alle de brieven* 1, 30.
33 In later years, he may have received help from Adriaan Jansz Lansvelt: 'Biografisch register', *Alle de brieven*, vol. IV.
34 Letter no. 8 [4.], 1 June 1674, *Alle de brieven* 1, 114. This reference was noted by Brian J. Ford, though he suggests that it is in the first letter. Brian J. Ford, 'Antoni van Leeuwenhoek – microscopist and visionary scientist', *Journal of Biological Education* 23, 4 (1989), 293–9, in particular 295.
35 Brian J. Ford, *The Leeuwenhoek Legacy* (Bristol, 1991), 51–8.
36 Hooke, *Micrographia*, 115.
37 In 1682 Antoni was able to understand a letter from Robert Hooke, possibly with help. See letter no. 68 [36], 4 April 1682, *Alle de brieven* 3, 420, note 7. In 1694 he seemed to have understood a letter from Richard Waller without help: letter no. 138, 26 May 1694, *Alle de brieven* 10, 140. And in 1704, at least part of *Philosophical Transactions*: letter no. 248, 21 March 1704, *Alle de brieven* 14, 326. With thanks to Douglas Anderson.
38 Letter no. 4, *Alle de brieven* 1, 66.
39 Edward G. Ruestow, *The Microscope in the Dutch Republic: The Shaping of Discovery* (Cambridge and New York, 1996), 188.
40 Eric Jorink, 'Swammerdam, hoveling? Enige kanttekeningen bij de reputatie van een wetenschappelijk onderzoeker', *Studium* 8, 4 (2016), 178.
41 Jan Swammerdam, *Bybel der natuure door Jan Swammerdam, Amsteldammer, of historie der insecten*, vol. II (Leiden, 1737), 835.
42 Letter no. 9 [5] 6 July 1674, *Alle de brieven* 1, 122.
43 Letter no. 5 [3], 7 April 1674, *Alle de brieven* 1, 74.
44 Letter no. 80 [41], 14 April 1684, *Alle de brieven* 4, 240.
45 Letter no. 33 [21], 5 October 1677, *Alle de brieven* 2, 242.
46 Ibid.
47 Ibid., 244.
48 Ibid., 244.
49 Letter no. 37 [23], 14 January 1678, *Alle de brieven* 2, 306.
50 Letter no. 67 [35], 3 March 1682, *Alle de brieven* 3, 404; letter no. 80 [41], 240, 242; letter no. 214 [128], 9 July 1700, *Alle de brieven* 13, 148.
51 Letter no. 37 [23], 306.
52 Letter no. 214 [128], 142.
53 Ruestow, *The Microscope in the Dutch Republic*, 194.

54 One disadvantage is that blood cells burst in water but, strangely enough, Antoni did not appear to have experienced that problem.

55 Ruestow, *The Microscope in the Dutch Republic*, 194; Antoni van Leeuwenhoek, *Send-Brieven zoo aan de Hoog Edele Heeren van de Koninklyke Societeit te Londen, als aan andere aansienelyke en geleerde lieden, over verscheyde verborgentheden der natuure* (Delft, 1718), 435, 436; letter no. 349, 8 October 1717, *Alle de brieven* 18.

56 Letter no. 37 [23], 308; letter no. 65 [33], 12 November 1680, *Alle de brieven* 3, 286. Antoni was not at first sure of the structure of the flat globules. In the blood globules of cod and salmon, he saw '3, 4, 5, 6, nay as many as 8' smaller globules: letter no. 67 [35], 406. But later, he decided after examining the blood of a flounder, that the oval particles consisted of six globules: letter no. 214 [128], 144.

57 Letter no. 62 [32], 14 June 1680, *Alle de brieven* 3, 254.

58 Ibid.

59 Letter no. 37 [23], 314–16; letter no. 40 [26], 27 September 1678, *Alle de brieven* 2, 388.

60 Letter no. 37 [23], 312.

61 Letter no. 70 [37], 22 January 1683, *Alle de brieven* 4, 38.

62 Ruestow, *The Microscope in the Dutch Republic*, 190–91.

63 Letter no. 62 [32], 244–50.

64 Letter no. 65 [33], 282.

65 Letter no. 376, August 1723, *Alle de brieven* 19.

66 Letter no. 339 [XXXIV], 6 March 1717, *Alle de brieven* 18.

67 Ruestow, *The Microscope in the Dutch Republic*, 189.

68 Letter no. 52, 120, 122.

69 Letter no. 82 [43], 5 January 1685, *Alle de brieven* 5, 62, 64.

70 Tuhina Neogi et al., 'Alcohol quantity and type on risk of recurrent gout attacks: An internet-based case-crossover study', *American Journal of Medicine* 127, 4 (2014), 311–18.

4: Mind-Bogglingly Multitudinous

1 Antoni bequeathed a copper quadrant, according to the inventory of Het Gouden Hoofd after Maria's death: 'Inventaris van de boedel die Juffrouw Maria Van Leeuwenhoek . . .', in 'Minuutakten Joris Geesteranus januari–november 1745', Delft Municipal Archives, 161-2791, 90–122v, in particular 108v.

2 Letter no. 299 [IV], 14 March 1713, *Alle de brieven* 17, 84.

3 See for example Mattheus van Nispen, *De beknopte lant-meet-konst: Leerende in t' korte, alles wat in t' gemeen, in de practijcke des landt-metens voor-komen kan* (Dordrecht, 1662), 245–8.

4 A. Schierbeek, 'Een paar nieuwe bijzonderheden over Van Leeuwenhoek', *Nederlands Tijdschrift voor Geneeskunde* 74 (August 1930), 3891–3.

5 Harry Lintsen, ed., *Geschiedenis van de techniek in Nederland 5: Techniek, beroep en praktijk* (The Hague, 1994), 26.

6 Douglas Anderson, 'Appointed co-wine gauger to help Dirk Arisz', *Lens on Leeuwenhoek*, https://lensonleeuwenhoek.net, accessed 19 December 2022; A. Schierbeek, *Antoni van Leeuwenhoek. Zijn leven en zijn werken* (Lochem, 1950), 25–6.

Wine gaugers worked with measuring sticks, and the inventory of Het Gouden Hoofd after Maria's death shows that Antoni left a collection of these sticks. The inventory lists: two black ebony measuring sticks with silver knobs and rings, one ditto silver-mounted on both sides, one ditto silver-mounted on one side, miscellaneous ebony measuring sticks: 'Inventaris van de boedel die Juffrouw Maria Van Leeuwenhoek . . .', in 'Minuutakten Joris Geesteranus januari–november 1745', Delft Municipal Archives, 161-2791, 90–122v, in particular 108v.

7 Dirck Evertsz van Bleijswijck et al., *Kaart Figuratief* (Delft, 1678), Delft Municipal Archives.

8 Douglas Anderson, 'St. Nicolaas Gilde', *Lens on Leeuwenhoek*, https://lensonleeuwenhoek.net, accessed 19 October 2022.

9 A sister of Antoni's mother, Cornelia Jacobs van den Berch, was married to Simon Elsevier (or Sijmon Elsevyer): 'Ondertrouwboeken Oude Kerk', Delft Municipal Archives, 14-21, 58. Their daughter Aletta (or Alette) was baptized on 27 August 1641: 'Doopboeken Oude Kerk, Delft, 1 augustus 1624–31 juli 1642', Delft Municipal Archives, 14-8, 213v. On 25 March 1638, Jacob and Lijsbeth Spoors baptized their son Johan (or Johannis): 'Doopboeken Nieuwe Kerk 14 augustus 1636–1649', Delft Municipal Archives, 14-56, 20v. Aletta and Johan announced their banns on 20 September 1659: 'Ondertrouwboeken Oude Kerk', Delft Municipal Archives, 14-22, 53. They baptized their son Sijmon in the Nieuwe Kerk on 31 July 1667: 'Doopboeken Nieuwe Kerk, 1661–1682', Delft Municipal Archives, 14-58, 89. They apparently buried two children before that, in 1665 and 1666. See also: Douglas Anderson, 'Jacob Spoors', *Lens on Leeuwenhoek*, https://lensonleeuwenhoek.net, accessed 22 December 2022.

10 Huib J. Zuidervaart and Marlise Rijks, '"Most rare workmen": Optical practitioners in early seventeenth-century Delft', *British Journal for the History of Science* 48, 1 (2015), 53–4.

11 Spoors observed the candle from a distance of '800 *Rynlansche roeden*' and one *roede* was around 3.77 metres (2.34 yds) long. Isaac Beeckman and Cornelis de Waard, eds, *Journal tenu par Isaac Beeckman de 1604 à 1634*, vol. III (The Hague, 1945), 321.

12 Jacob Spoors, *Oratie van de nieuwe wonderen des wereldts, de nuttigheyd, de waerdigheyd, der wis- ende meetkunsten* (Delft, 1638), 6.

13 Zuidervaart and Rijks, 'Most rare workmen', 65; Spoors, *Oratie van de nieuwe wonderen des wereldts*.

14 Letter no. 11 [6], 7 September 1674, *Alle de brieven* 1, 162.

15 Ibid., 164.

16 Jacob Spoors, *'Caart ende ontworp gedaan bij mijn Jacob Spoors geadmitteert landtmeter bijden Hove van Hollandt tot Delft residerende vanhet noordteinde van Berckel met affteickeninge van een geconcipieerde nieuwe te make vaart ofte watering, vande Berckelse kade aff in het noorde ende soo voorts zuijdwaarts op tot inde Oostermeer van Berckel, sooals die op de kaart affgeteickent is, doch alles op approbatie vande wel edele hoge heemraden van Delfflandt'* (1675); Anderson, 'Jacob Spoors', accessed 22 December 2022.

17 'Plan van David Coornwinder voor de droogmaking van het Oostmeer in de polder van Berkel', 1641, Rotterdam Municipal Archives, 1305-88.

18 Letter no. 11 [6], 162.

19 With thanks to Wim van Egmond, who investigated this subject.

20 Letter no. 47, 20 May 1679, *Alle de brieven* 3, 58.

21 Letter no. L-557, 16 August 1717, *Alle de brieven* 18.

22 Letter no. 47, 56.

23 Ibid.

24 Letter no. 339 [XXXIV], 6 March 1717, *Alle de brieven* 18.

25 Letter no. 47, 88. Two years earlier, Antoni wrote briefly about minuscule blood vessels in the brain: 'I have seen in the Brain and most in the Cortical part, such Small sanguineous vessels being red (which came out of bigger ones) that I cannot comprehend, how the globuls could pass through them': letter no. 32 [20], 14 May 1677, *Alle de brieven* 2, 222.

26 Letter no. 47, 54.

27 Ibid., 58.

28 Letter no. 11 [6], 1, 164.

29 Letter no. 47, 58.

30 Ibid., 62.

31 Antoni also wrote that the vessels are 'more than a thousand times thinner than a hair of one's head' and of 'the little vessels in our body 1,089 thinner than a hair'. But from the context, it is clear that here he is talking about the surface area of a cross-section. When he talks about the width of a hair against the thickness, he states that as a ratio of 33 to 1 ($33^2 = 1{,}089$): letter no. 47, 358–60.

32 Antoni first calculated the medium diameter of his body if he were to be sliced through the middle above his hips. That cross-section would be approximately a circle and, as all present-day schoolchildren learn, the surface area of a circle = $\pi \times (\text{radius})^2$. Antoni, however, did not know of this formula. He wrote that Archimedes had proved 'that as 14 is to 11 so the square of the diameter is to the surface of a circle'. In slightly more modern terms that means that 14 : 11 = diameter^2: surface area. Or more familiarly, 14 : 11 = $(2 \times \text{radius})^2$: surface area. With a little reshuffling, we get: surface area = $22/7 \times (\text{radius})^2$. Antoni thus used 22/7 as an approximation for π. He then calculated that the surface area of the cross-section of his body was $22/7 \times (1/2 \times 8)^2 = 50\ 2/7$ square inches, which he rounded off to 50.

33 × 600 capillaries fitted into an inch, meaning that there were $(33 \times 600)^2$ in a square inch. Thus $50 \times (33 \times 600)^2$ fitted into the diameter of his waist = 19,602,000,000 diameters of the capillaries: letter no. 47, 58–62.

33 Ibid., 62.

34 Ibid., 64.

35 Ibid.

36 In his *History of the Royal Society of London*, Thomas Birch seems to suggest that Antoni had even applied himself to the fundamental (and, as we now know, unsolvable) problem of doubling the cube. In his book he recorded that, on 31 May 1677, the members 'Read Monr. LEEWENHOECK and Dr. WALLIS'S account on the duplication of cubes, and of the instant motion of light.' But it is more probable that, on that day, the members read letter 32 [20], which Antoni had sent on 14 May 1677 and in which he described a variety of observations, but said nothing about cubes. And that only mathematician John Wallis had discussed the doubling of the cube, as he had on previous occasions. Thomas Birch, *History of the Royal Society of London*, vol. III (London, 1757), 340; letter no. 32 [20].

37 'Biografisch register' in *Alle de brieven* 11, 336; Peter H. Meurer, 'Cornelis van Beughem, een vergeten randfiguur van de Amsterdamse kartografie', *Caert Thresoor. Tijdschrift voor de Geschiedenis van de Kartografie* 31, 2 (2012), 39–46.

38 Letter from Pieter Rabus, 16 May 1696, *Alle de brieven* 11, 260.

39 Ibid., 262.

40 Ibid., 260.

41 Ibid.

42 Jan Trioen et al., *Nodige verantwoordinge voor de heer Pieter Rabus en juffr. syne huysvrouw tegens de Amsterdammers en Haarlemmers, niet gelovende de werking der wichelroeden* (n.p.), 4, 7.

43 Balthasar Bekker, '4, Waar in 't bewijs, dat uit d'Ervarentheid genomen word, ten gronde toe word ondersocht', in *De betoverde wereld* (Amsterdam, 1691), 196.

44 Ibid.,, 202–4.

45 Letter from Pieter Rabus, 16 May 1696, 266.

46 Letter no. 166, 1 June 1696, *Alle de brieven* 11, 270.

47 Ibid., 270.

48 Ibid., 270, 272.

49 Letter no. 184 [108], 5 April 1697, *Alle de brieven* 12, 138.

50 Letter no. 166, 272.

51 Pieter Rabus, *Boekzaal van Europe*, May–June 1697, 422–3.

52 Trioen et al., *Nodige verantwoordinge voor de heer Pieter Rabus en juffr*, 4, 7.

53 Pieter Rabus, 'Afkeer eener lasterbende', *Boekzaal van Europe*, October 1697, 379–80.

54 Letter from Pieter Rabus, 30 July 1696, *Alle de brieven* 12, 24.

55 Letter no. 184 [108], 138.

56 Pieter Rabus, *Boekzaal van Europe*, January–February 1697, 68–72.
57 Ibid., 70–71.
58 Ibid., 72.
59 Ibid., 73–6.
60 Ibid., 76.
61 Herman Lufneu, 'Een brief, over de onmogelijkheid der zoo genaamde Sympathetische werking', in Rabus, *Boekzaal van Europe*, January–February 1697, 76–140.
62 Ibid., 125–30.
63 Ibid.
64 Rabus, *Boekzaal van Europe*, May–June 1697, 435.
65 Letter no. 184 [108], 154.
66 Ibid.
67 Ibid., 154, 156.
68 Ibid., 156.
69 Letter no. 47, 54.

5: Little Animals and the Most Addle-Pated Propositions among Physicians

1 Letter no. 38 [24], 18 March 1678, *Alle de brieven* 2, 338.
2 Ibid., 340.
3 Letter no. 39 [25], 31 May 1678, *Alle de brieven* 2, 360.
4 Ibid., 362.
5 In 1677 the request had been made three or four years earlier. Oldenburg wrote to Antoni from 1674: letter no. 35 [22], November 1677, *Alle de brieven* 2, 290.
6 Ibid., 290.
7 H. Halbertsma, 'Ontleedkundige aanteekeningen. VI. Johan Ham van Arnhem, de ontdekker der spermatozoïden', *Verslagen en Mededeelingen der Koninklijke Akademie van Wetenschappen. Afdeeling Natuurkunde* XIII (1862), 342.
8 Letter no. 196 [113], 17 December 1698, *Alle de brieven* 12, 254; letter no. 35 [22], 280, 282.
9 Letter no. 35 [22], 280, 282.
10 Ibid., 280–82.
11 Ibid., 284.
12 Ibid., 290.
13 'Ondertrouwboek Gerecht, 1670–1674' , Delft Municipal Archives, 14-131, 24.
14 W. H. van Seters, 'Van Leeuwenhoeks tweede huwelijk', *Nederlands Tijdschrift voor Geneeskunde* 112, 27 (1968), 1258.
15 Letter no. 35 [22], 284.
16 Ibid., 284, 286.

17 Ibid., 288.
18 Letter no. 35 [22], 280.
19 Letter no. 11 [6], 7 September 1674, *Alle de brieven* 1, 138–62.
20 Ibid., 162–4.
21 See for example letter no. 26 [18], 9 October 1676, *Alle de brieven* 2, 84, 122, 136; letter no. 37 [23], 14 January 1678, *Alle de brieven* 2, 320.
22 Letter no. 19 [13], 20 December 1675, *Alle de brieven* 1, 330; letter no. 20, 22 January 1676, *Alle de brieven* 1, 346.
23 Letter no. 26 [18], 64. In this letter, Antoni described a study he conducted 'in the year 1675 about mid-September'. He was probably referring to the experiment described here: letter no. 2, 15 August 1673, *Alle de brieven* 1, 58.
24 Letter no. 26 [18], 66.
25 Ibid., 66–8.
26 Ibid., 72–88.
27 Ibid., 74, 94.
28 Ibid., 84.
29 Ibid., 64.
30 Ibid., 134.
31 Ibid., 144.
32 Clifford Dobell, *Antony van Leeuwenhoek and His 'Little Animals': Being Some Account of the Father of Protozoology and Bacteriology and His Multifarious Discoveries in These Disciplines*, new edn (New York, 1960).
33 Ibid., 118, note 2.
34 Ibid., 118, note 3.
35 Wim van Egmond, 'The riddle of the "Green Streaks"', *Micscape*, no. 239 (2016); letter no. 11 [6], 162.
36 Letter no. 35 [22], 292.
37 Letter no. 38 [24], 326.
38 Letter no. 35 [22], 292–4.
39 Letter no. 38 [24], 336, 338.
40 Ibid., 328, 336, 340.
41 Ibid., 362.
42 'MicroLab3: Bodily Fluids', *Visualizing the Unknown*, https://visualizingtheunknown.com, accessed 9 November 2022.
43 Letter no. 38 [24], 328.
44 Letter no. 39 [25], 362.
45 Letter no. 43 [28], 25 April 1679, *Alle de brieven* 3, 12–14.
46 Ibid., 14–16.
47 Ibid., 18.
48 Ibid., 20.
49 Ibid.
50 Letter no. 70 [37], 22 January 1683, *Alle de brieven* 4, 10; Matthew Cobb, *Generation: The Seventeenth-Century Scientists Who Unraveled the Secrets of Sex, Life, and Growth*, 1st U.S. edn (New York, 2006), 205.

51 Letter no. 57 [30], 5 April 1680, *Alle de brieven* 3, 204; letter no. 200 [116] 9 June 1699, *Alle de brieven* 12, 304; Edward G. Ruestow, *The Microscope in the Dutch Republic: The Shaping of Discovery* (Cambridge and New York, 1996), 250–53. One of the problems with this theory was that every man would carry the seed of all his future generations of descendants.
52 Letter no. 84 [45], 30 March 1685, *Alle de brieven* 5, 184; letter no. 200 [116], 298.
53 Letter no. 220 [135], 25 December 1700, *Alle de brieven* 13, 240.
54 Letter no. 38 [24], 332.
55 Luuc Kooijmans, *Gevaarlijke kennis: Inzicht en angst in de dagen van Jan Swammerdam* (Amsterdam, 2007), 103–5, 118, 133, 165.
56 Van Horne and Swammerdam did their research in 1667; De Graaf started in 1671. Cobb, *Generation*, 161, 165.
57 Ibid., 74–8.
58 Letter no. 38 [24], 328–32.
59 Ibid., 332–4.
60 Letter no. 84 [45], 152.
61 Ibid., 152, 154.
62 Ibid., 154, 156.
63 Letter no. 38 [24], 342–4.
64 Letter no. 70 [37], 6.
65 Letter no. 38 [24], 344.
66 Ibid., 344–6.
67 Letter no. 39 [25], 358.
68 Letter no. 70 [37], 8.
69 Ibid., 6.
70 Cobb, *Generation*, 212; letter no. 70 [37], 6.
71 Letter no. 70 [37], 12.
72 Ibid., 14.
73 Letter no. 84 [45], 174.
74 Ibid., 176.
75 Cobb, *Generation*, 211, 212.
76 Letter no. 135 [81], 19 March 1694, *Alle de brieven* 10, 54–6.
77 Ibid., 56, 58.
78 Letter no. L-588, August 1723, *Alle de brieven* 19; Cobb, *Generation*, 214.
79 According to Hartsoeker himself, that visit took place around the end of 1672, beginning of 1673, so before Antoni had achieved international fame through the Royal Society. That would have been remarkably early. Nicolaas Hartsoeker, 'Extrait critique des lettres de feu M. Leeuwenhoek', in *Cours de physique accompagné de plusieurs piéces concernant la physique qui ont déja paru et d'un extrait critique des lettres de M. Leeuwenhoek* (The Hague, 1730), 7. Antoni wrote long after the visit, in 1698, that Hartsoeker was then a young student (letter no. 196 [113], 252). If that was true, the visit could not have occurred earlier than 1674, when according to Hartsoeker

he began attending the Athenaeum Illustre in Amsterdam. But it is also possible that, half a century later, Antoni was mistaken and Hartsoeker had not begun his studies at the time of his visit.

80 Hartsoeker, 'Extrait critique des lettres de feu M. Leeuwenhoek', 43; letter no. 196 [113], 252.

81 Hartsoeker, 'Extrait critique des lettres de feu M. Leeuwenhoek', 44.

82 Ibid., 45.

83 Letter no. 35 [22], 280.

84 Hartsoeker, 'Extrait critique des lettres de feu M. Leeuwenhoek', 45, 46.

85 Ibid., 46.

86 Ibid., 1–66. The essay was published after both had died.

87 Ibid., 46.

88 Letter no. 8 [4], 1 June 1674, *Alle de brieven* 1, 110.

89 Letter no. 42 [27], 21 February 1679, *Alle de brieven* 2, 412.

90 Letter no. 76 [39], 17 September 1683, *Alle de brieven* 4, 124; letter no. 123 [75], 16 September 1692, *Alle de brieven* 9, 132.

91 Letter no. 76 [39], 124.

92 Ibid., 124, 126.

93 'No. 2117, Nicolaas Hartsoeker à Christiaan Huygens, 14 mars 1678', in *Oeuvres complètes*, vol. III: *Correspondance, 1676–84* (The Hague, 1899), 58–61. Twelve days later, on 26 March 1678, Christiaan Huygens wrote to his brother Constantijn on sperm research, mentioning a letter from Antoni on the subject. Francis Cole has interpreted that as an indication that Huygens obtained his knowledge of sperm cells from Van Leeuwenhoek (Francis Cole, *Early Theories of Sexual Generation* (Oxford, 1930), 2).

Hartsoeker also had his information from Antoni, according to Cole, perhaps via Huygens. Cole assumed that, in his letter to his brother, Christiaan Huygens referred to Antoni's first letter on little seed animals of November 1677. But that cannot be correct, as Huygens wrote to his brother that Van Leeuwenhoek had counted a million little seed animals in a drop of semen the size of a grain of sand. Van Leeuwenhoek did not do that in his letter of November 1677. There, he spoke only of more than a thousand little animals in a grain of sand (letter no. 35 [22], 284). The number of a million (or, in Antoni's words, 'ten hundreds of thousands') is stated in a later letter, no. 38 [24], 18 March 1678, 348. That letter thus dates from after Hartsoeker's letter to Huygens on sperm does not prove that Huygens got his knowledge from Antoni in the first instance. And perhaps more importantly, the letter does not prove that Hartsoeker knew of Antoni's discovery before he spoke and wrote to Huygens on the little seed animals.

94 Christiaan Huygens, 'Extrait d'une Lettre de M. Huguens de l'Acad. R. des Sciences à l'Auteur du Journal, touchant une nouvelle maniere de Microscope qu'il a apporté de Hollande', *Journal des sçavans* (15 August 1678), 345–7.

95 Hartsoeker, 'Extrait critique des lettres de feu M. Leeuwenhoek', 46.

96 Ibid.

97 Christiaan Huygens wrote about sperm '*cujus primus auctor Hammius quidam perhibetur*' (of which it is said that the first to discover it was a certain Ham). 'No. 2125. Christiaan Huygens à N. Grew, 6 juin 1678', *Oeuvres complètes*, vol. III, 77. In his history of the Royal Society, Thomas Birch translated this as, 'the discovery of those animalcules in semine animali made, by one Hammius, a student at Leyden'. Thomas Birch, *The History of the Royal Society of London*, 4 vols (London, 1757), vol. III, 415. But that was more assertive than Huygens's original sentence.
98 Hartsoeker, 'Extrait critique des lettres de feu M. Leeuwenhoek', 7.
99 Nicolaas Hartsoeker, *Proeve der deurzicht-kunde*, translated into Dutch by A. Block (Amsterdam, 1699), 223. The title page states the year 1699, but Antoni read the book in 1698.
100 Letter no. 220 [135], 232.
101 Letter no. 196 [113], 252, 254.
102 Nicolaas Hartsoeker, *Essay de dioptrique* (Paris, 1694), 227.
103 Hartsoeker, 'Extrait critique des lettres de feu M. Leeuwenhoek', 1–66.

6: Openness and Smokescreens

1 Letter no. 1 [1], 28 April 1673, *Alle de brieven* 1, 32.
2 Letter no. 26 [18], 9 October 1676, *Alle de brieven* 2, 78–80; letter no. 84 [45], 30 March 1685, *Alle de brieven* 5, 1685–6, 200; letter no. 133 [79], 24 February 1694, *Alle de brieven* 9, 366; letter no. 200 [116], 9 June 1699, *Alle de brieven* 12, 308.
3 Douglas Anderson, 'The tensions between facts and fantasy', *Studium* 8, 4 (24 May 2016), 230; Marlies Philippa, Frans Debrabandere and Arend Quak, *Etymologisch woordenboek van het Nederlands* (Amsterdam, 2005), 628–9. Reinier Boitet also used the word *komptoir* for the office in William Davidson's company where Antoni gained experience as a bookkeeper. Reinier Boitet, *Beschryving der stadt Delft, behelzende een zeer naaukeurige en uitvoerige verhandeling van deszelfs eerste oorsprong, benaming, bevolking, aanwas, gelegenheid, prachtige en kunstige gedenkstukken en zeltzaamheden. Nevens derzelver voorregten, handvesten, previlegien, en regeeringsvorm: alles 't zamengestelt en getrokken uit oude handtschriften, memorien, en brieven, en met zeer veele echte bewysstukken (te vooren noit gedrukt) bevestigt: door verscheide liefhebbers en kenners der Nederlandsche oudheden* (Delft, 1729), 765.
4 Douglas Anderson, '"Your Most Humble Servant": The letters of Antony van Leeuwenhoek', *FEMS Microbiology Letters* 369, 1 (4 March 2022), 5.
5 Letter no. 167 [100], 6 July 1696, *Alle de brieven* 11, 278–86; letter no. 197 [114], 1 February 1699, *Alle de brieven* 12, 270–76; Huib J. Zuidervaart and Douglas Anderson, 'Antony van Leeuwenhoek's microscopes and other scientific instruments: New information from the Delft archives', *Annals of Science* 73, 3 (2016), 266, note 67.

6 Letter 113 [66], 12 January 1689, *Alle de brieven* 8, 114, 166.
7 Douglas Anderson, 'Still going strong: Leeuwenhoek at eighty', *Antonie van Leeuwenhoek. Journal of Microbiology* no. 106 (2014), 3–26, in particular 14.
8 Zacharias Conrad von Uffenbach, *Merkwürdige Reisen durch Niedersachsen, Holland und Engelland*, vol. III (Ulm, 1754), 360.
9 Ibid., 358.
10 *Catalogus van het vermaarde cabinet van vergrootglasen, met zeer veel moeite, en kosten in veele jaren geïnventeert, gemaakt, en nagelaten door wylen den heer Anthony van Leeuwenhoek, in zyn Ed: Leeven Lid van de Koninglyke Societeit der Wetenschappen in Londen* (Delft, 1747).
11 Nicolaas Hartsoeker, 'Extrait critique des lettres de feu M. Leeuwenhoek', in *Cours de physique accompagné de plusieurs piéces concernant la physique qui ont déja paru et d'un extrait critique des lettres de M. Leeuwenhoek* (The Hague, 1730), 7. The first time that Antoni wrote about the little seed animals of fleas was in letter no. 65 [33] of 12 November 1680, so a long time after Hartsoeker's visit, which was reported to be in 1679. It is therefore possible that Hartsoeker's later account of the meeting was not entirely accurate.
12 Hartsoeker, 'Extrait critique des lettres de feu M. Leeuwenhoek', 8.
13 Ibid.
14 Thomas Birch, *The History of the Royal Society of London*, 4 vols (London, 1757), vol. IV, 365; Peter King, *The Life of John Locke, with Extracts from His Correspondence, Journals, and Common-Place Books*, vol. I (London, 1830), 309.
15 Von Uffenbach, *Merkwürdige Reisen*, vol. III, 359.
16 Tiemen Cocquyt et al., 'Neutron tomography of Van Leeuwenhoek's microscopes', *Science Advances* 7, 20 (2021).
17 Ibid.
18 Robert Hooke, *Micrographia: or Some Physiological Descriptions of Minute Bodies Made By Magnifying Glasses With Observations and Enquiries thereupon* (London, 1665), preface.
19 Hartsoeker, 'Extrait critique des lettres de feu M. Leeuwenhoek', 44; no. 2118, N. Hartsoeker à Christiaan Huygens, 25 mars 1678, in Christiaan Huygens, *Oeuvres complètes*, vol. III: *Correspondance, 1676–84* (The Hague, 1899).
20 Huib J. Zuidervaart and Douglas Anderson, 'Antony van Leeuwenhoek's microscopes and other scientific instruments: New information from the Delft archives', *Annals of Science* 73, 3 (2016), 261.
21 J. van Zuylen, 'The microscopes of Antoni van Leeuwenhoek', in *Antoni van Leeuwenhoek 1632–1723: Studies on the Life and Work of the Delft Scientist Commemorating the 350th Anniversary of His Birthday*, ed. L. C. Palm and H.A.M. Snelders (Amsterdam, 1982), 29–55, in particular 43–6.
22 Reinhold Reith, 'Circulation of skilled labour in late medieval and early modern Central Europe', in *Guilds, Innovation and the European Economy, 1400–1800*, ed. S. R. Epstein and Maarten Prak (Cambridge, 2008), 114–42, 136.

23 Pamela H. Smith, 'What is a secret? Secrets and craft knowledge in early modern Europe', in *Secrets and Knowledge in Medicine and Science, 1500–1800*, ed. Elaine Yuen Tien Leong and Alisha Michelle Rankin (Farnham, UK and Burlington, VT, 2011), 47–66, in particular 49.
24 Hans Bots, *De Republiek der Letteren: De Europese intellectuele wereld 1500–1760* (Nijmegen, 2018), 55.
25 Anne Goldgar, *Impolite Learning: Conduct and Community in the Republic of Letters, 1680–1750* (New Haven, CT, 1995), 22; Bots, *De Republiek der Letteren*, 56–7.
26 Barbara J. Shapiro, 'Testimony in seventeenth-century English natural philosophy: Legal origins and early development', *Studies in History and Philosophy of Science* 33 (2002), 250–51.
27 Letter no. 15 [9], 22 January 1675, *Alle de brieven* 1, 210.
28 Letter no. 32 [20], 14 May 1677, *Alle de brieven* 2, 216; letter no. 37 [23], 14 January 1678, *Alle de brieven* 2, 314; letter no. 82 [43], 5 January 1685. *Alle de brieven* 5, 30.
29 Tiemen Cocquyt et al., 'Visualizing the unknown: Survey and themes', article under preparation.
30 Letter no. 2, 15 August 1673, *Alle de brieven* 1, 42.
31 Sietske Fransen, 'Antoni van Leeuwenhoek, his images and draughtsmen', *Perspectives on Science* 27, 3 (2019), 506–8.
32 Ibid., 511.
33 Ibid., 528.
34 Ibid., 522; Boitet, *Beschryving der stadt Delft*, 791.
35 Fransen, 'Antoni van Leeuwenhoek, his images and draughtsmen', 524.
36 Ibid., 530.
37 Letter no. 13 [8], 4 December 1674, 476.
38 Letter no. 33 [21], 5 October 1677, *Alle de brieven* 2, 246.
39 Letter no. 101 [56], 11 July 1687, *Alle de brieven* 6, 318.
40 Letter no. 164 [98], 20 February 1696, *Alle de brieven* 11, 188, 190.
41 Ibid., 190–94.
42 Ibid., 192–4.
43 Letter no. 156, 10 September 1695, *Alle de brieven* 11, 80.
44 Letter no. 8 [4], 1 June 1674, *Alle de brieven* 1, 96; letter no. 9 [5], 6 July 1674, *Alle de brieven* 1, 122.
45 Letter no. 8 [4], 96.
46 This became clear to see during the observations made by the research project Visualizing the Unknown.
47 Letter no. 8 [4], 108.
48 Letter no. 9 [5], 118.
49 Letter no. 15 [9], 210.
50 Letter no. 17 [11], 26 March 1675, *Alle de brieven* 1, 292.
51 Letter no. 8 [4], 114.
52 Letter no. 200 [116], 9 June 1699, *Alle de brieven* 12, 308.

53 Letter no. 26 [18], 80.
54 Letter no. 200 [116], 306–8.
55 Letter no. 236 [146], 20 April 1702, *Alle de brieven*, 106.
56 Anderson, 'Still going strong: Leeuwenhoek at eighty', 15; letter no. 307, 21 August 1714, *Alle de brieven* 17, 200.
57 Letter no. 200 [116], 296.
58 Ibid., 304.
59 P. van der Lijn, *Nederlandse zwerfstenen* (Zutphen, 1935), 82; *Catalogus van het vermaarde cabinet van vergrootglasen*, 31, 33, no. 110, 117, 123.
60 *Catalogus van het vermaarde cabinet van vergrootglasen*, 33, no. 126.
61 Cocquyt et al., 'Neutron tomography of Van Leeuwenshoek's microscopes'.
62 Ibid.
63 Ibid.
64 Smith, 'What is a secret?', 49.
65 Letter no. 200 [116], 296.

7: Strange Curiosities and Senseless Details

1 Letter no. 80 [41], 14 April 1684, *Alle de brieven* 4, 244, 248.
2 Ibid., 48.
3 Tim van Polanen, 'Snak, Claas and Bastiaan's struggle for freedom: Three Curaçaoan enslaved men and their court cases about the free soil principle in the Dutch Republic', BMGN – *Low Countries Historical Review* 136, 1 (2021), 34.
4 Letter no. 80 [41], 248–50.
5 Ibid., 246.
6 Ibid., 246, 248.
7 *Catalogus van het vermaarde cabinet van vergrootglasen, met zeer veel moeite, en kosten in veele jaren geïnventeert, gemaakt, en nagelaten door wylen den heer Anthony van Leeuwenhoek, in zyn Ed: Leeven Lid van de Koninglyke Societeit der Wetenschappen in Londen* (Delft, 1747), 15.
8 Ibid., 17.
9 Ibid., 27; Gertrud Konings, 'Opuntia Ficus-Indica as an extremophile: A survivor in extreme climates?', *Cactus and Succulent Journal* 82, 4 (2010), 172–5.
10 *Catalogus van het vermaarde cabinet van vergrootglasen*, 31.
11 Roelof van Gelder, 'De wereld binnen handbereik', in *De wereld binnen handbereik: Nederlandse kunst- en rariteitenverzamelingen, 1585–1735*, ed. Ellinoor Bergvelt and René Kistemaker (Zwolle and Amsterdam, 1992), 15–38, in particular 15–16.
12 Antoni's collected letters, with illustrations, can also be seen as his version of a curiosity cabinet: Douglas Anderson, 'Leeuwenhoek's Cabinet of Wonders', *Lens on Leeuwenhoek*, https://lensonleeuwenhoek.net, accessed 6 December 2022.

13 Letter no. L-539 [XXVIII], 28 September 1716, *Alle de brieven* 18.
14 Letter no. L-567, 24 January 1721, *Alle de brieven* 19.
15 Letter no. 32 [20], 14 May 1677, *Alle de brieven* 2, 228.
16 Ibid., 228–30.
17 Ibid. In reality, moxa also contains leaves, for example from mugwort.
18 Ellinoor Bergvelt, Debora J. Meijers and Alexis Joachimides, eds, *Kabinetten, galerijen en musea: Het verzamelen en presenteren van naturalia en kunst van 1500 tot heden*, revd edn (Zwolle, 2013), 169–70.
19 Nicolaas Hartsoeker, *Eclairissement sur les conjectures physiques* (Amsterdam, 1710), 82, 83.
20 S. Catherine Abou-Nemeh, 'The natural philosopher and the microscope: Nicolas Hartsoeker unravels nature's "Admirable OEconomy"', *History of Science* 51, 1 (March 2013), 1–32, in particular 11, 12.
21 Letter no. 80 [41], 244, 246.
22 Letter no. 32 [20], 230, 232.
23 Marian Fournier, *The Fabric of Life: The Rise and Decline of Seventeenth-Century Microscopy* (Enschede, 1991), 122.
24 Letter no. 2, 15 August 1673, *Alle de brieven* 1, 46, 48; letter no. L-567.
25 P. Baas, 'Leeuwenhoek's contributions to wood anatomy and his ideas on sap transport in plants', in *Antoni van Leeuwenhoek 1632–1723: Studies on the Life and Work of the Delft Scientist Commemorating the 350th Anniversary of His Birthday*, ed. L. C. Palm and H.A.M. Snelders (Amsterdam, 1982), 79–107.
26 Letter no. 65 [33], 12 November 1680, *Alle de brieven* 3, 324.
27 Karen R. Zwier, 'Methodology in Aristotle's theory of spontaneous generation', *Journal of the History of Biology* 51, 2 (2018), 359, 373.
28 Hiro Hirai, 'Kircher's chymical interpretation of the creation and spontaneous generation', in Lawrence Principe, ed., *Chymists and Chymistry: Studies in the History of Alchemy and Early Modern Chemistry* (Sagamore Beach, MA, 2007), 77–8, in particular 78–9.
29 Ibid., 80–81.
30 Ibid., 83–5.
31 Ibid., 86.
32 Emily C. Parke, 'Flies from meat and wasps from trees: Reevaluating Francesco Redi's spontaneous generation experiments', *Studies in History and Philosophy of Science Part C: Studies in History and Philosophy of Biological and Biomedical Sciences* 45 (2014), 34–42, in particular 36.
33 Eric Jorink, 'De profeet en de boekhouder: Johannes Swammerdam, Antoni van Leeuwenhoek en de begindagen van de microscopie', in Johan Braeckman and Linda Van Speybroeck, eds, *Fascinerend leven: een geschiedenis van de biologie* (Gent, 2021), 179–205, in particular 185.
34 Ibid., 189.
35 Letter no. 11 [6], 7 September 1674, *Alle de brieven* 1, 142.
36 Ibid., 142, 144.

37 Letter no. 135 [81], 19 March 1694, *Alle de brieven* 10, 58.
38 Jorink, 'De profeet en de boekhouder', 189; 'Jan Swammerdam aan Melchisedec Thévento, 28 *April* 1678', in Jan Swammerdam, Melchisédech Thévenot and Gerrit Arie Lindeboom, *The Letters of Jan Swammerdam to Melchisedec Thévenot: With English Translation and a Biographical Sketch* (Amsterdam, 1975), 106.
39 Edward G. Ruestow, *The Microscope in the Dutch Republic: The Shaping of Discovery* (Cambridge and New York, 1996), 216, 218. As early as June 1679, Antoni asserted that all animals were created through reproduction, but that was in the context of his argument against the preformation theory, which postulated that later generations of young animals are present in miniature form in earlier generations: letter no. 49, 13 June 1679, *Alle de brieven* 3, 76. He wrote something similar in November 1680, but this time his comments were clearly intended to dispute the notion of spontaneous generation: letter no. 65 [33], 328.
40 Ruestow, *The Microscope in the Dutch Republic*, 203.
41 Ibid., 207.
42 Lodewijk Palm, 'Antoni van Leeuwenhoek's malacological researches as an example of his biological studies', in Palm and Snelders, *Antoni van Leeuwenhoek 1632–1723*, 152–67, in particular 155–6; letter no. 137 [83], 30 April 1694, *Alle de brieven* 10, 102.
43 Letter no. 137 [83], 92.
44 Alessandro Becchi, 'Between learned science and technical knowledge: Leibniz, Leeuwenhoek and the school for microscopists', in *Tercentenary Essays on the Philosophy and Science of Leibniz*, ed. Lloyd Strickland, Erik Vynckier and Julia Weckend (New York, 2016), 47–79, in particular 50, 59; letter no. 316 van Gottfried Wilhelm Leibniz, *Alle de brieven* 17.
45 Letter no. 322 [XX], 3 March 1716, *Alle de brieven* 17, 400.

8: Busy Times in Het Gouden Hoofd

1 Gerard van Loon, *Beschryving der Nederlandsche historiepenningen. Of beknopt verhaal van 't gene sedert de overdracht der heerschappye van keyzer Karel den Vyfden op koning Philips zynen zoon, tot het sluyten van den Uytrechtschen vreede, in de zeventien Nederlandsche gewesten is voorgevallen*, vol. IV (The Hague, 1731), 222–3.
2 Ibid., 223.
3 Ibid.
4 Ibid.
5 Letter no. 282, 14 January 1710, *Alle de brieven* 16.
6 Van Loon, *Beschryving der Nederlandsche historiepenningen*, 223.
7 Ibid.
8 Robert Hooke to Antoni van Leeuwenhoek, letter no. L-07218, 28 April 1678, *Alle de Brieven* 20.

9 Letter no. 26, 64–90.
10 Ibid., 94.
11 Letter no. 47, 20 May 1679, *Alle de brieven* 3, 56.
12 Letter no. 26, 94.
13 Antoni van Leeuwenhoek, 'Observations, communicated to the publisher by Mr. Antoni van Leewenhoeck, in a Dutch Letter of the 9th Octob. 1676. Here English'd: concerning little Animals by Him Observed in rain-well-sea-and snow water; as also in water wherein pepper had lain infused', *Philosophical Transactions* 12, 133 (25 March 1677), 821–31, in particular 827–31.
14 Letter no. 33 [21], 5 October 1677, *Alle de brieven*.
15 Ibid., 256–60.
16 Ibid., 262–70; 'Biografisch register' in *Alle de brieven* 2, 434–72, in particular 454.
17 Steven Shapin, *A Social History of Truth: Civility and Science in Seventeenth-Century England* (Chicago, IL, 1994), 123–5.
18 Thomas Birch, *History of the Royal Society of London*, 4 vols (London, 1757), vol. III, 347.
19 Ibid., 347.
20 Ibid., 352.
21 Robert Hooke to Antoni van Leeuwenhoek, letter no. L-07218, 28 April 1678, *Alle de brieven* 20.
22 Van Loon, *Beschryving der Nederlandsche historiepenningen*, 223.
23 Letter no. 51, 13 October 1679, *Alle de brieven* 3, 106, 108; letter no. 102 [57], 6 August 1687, *Alle de brieven* 7, 38, 40.
24 Letter no. 124, 23 September 1692, *Alle de brieven* 9, 172, note 9.
25 Letter no. 153, 16 August 1695, *Alle de brieven* 11, 51, note 5.
26 Antoni van Leeuwenhoek, *Send-Brieven zoo aan de Hoog Edele Heeren van de Koninklyke Societeit te Londen, als aan andere aansienelyke en geleerde lieden, over verscheyde verborgentheden der natuure* (Delft, 1718), Opdragt.
27 Van Leeuwenhoek, *Send-Brieven*, Opdragt.
28 Letter no. 102 [57], 6 August 1687, *Alle de brieven* 7, 38.
29 Leibniz to Henry Oldenburg, 18/28 November 1676, in *Sämtliche Schriften und Briefe. Dritte Reihe. Mathematischer, naturwissenschaftlicher und technischer Briefwechsel*, vol. V: *1691–3* (Berlin, 2003); Peter King, *The Life of John Locke, With Extracts From His Correspondence, Journals, and Common-Place Books*, vol. I (London, 1830), 309.
30 Letter no. 84 [45], 30 March 1685, *Alle de brieven* 5, 208; Cornelis Bontekoe, *Tractaat van het excellentste kruyd thee* (The Hague, 1678).
31 Letter no. 339 [XXXIV], 6 March 1717, Alle de brieven, 18.
32 Letter no. 72 [38], 16 July 1683, *Alle de brieven* 4, 96.
33 Cornelis Bontekoe, *Reden over de koortzen; Door welke aangewesen word, dat de gemene theorie en praktijk valsch, schadelijk en moordadig is* (The Hague, 1682), 54–5.

34 Letter no. 72 [38], 96.
35 Ibid., 96; Bontekoe, *Reden over de koortzen*, 16–17.
36 Douglas Anderson, 'Still going strong: Leeuwenhoek at eighty', *Antonie van Leeuwenhoek. Journal of Microbiology*, 106 (2014), 21.
37 Ibid.
38 Antoni van Leeuwenhoek, *Ondervindingen en beschouwingen der onsigtbare geschapene waarheden, waar in gehandelt wert vande schobbens inde mond, de lasarie, de jeuking, 't kind met vis-schobbens, 't binnenste der darmen, en de beweging derselve, als mede het vet dat inde selve gevonden wert: Geschreven aande wyt-beroemde Koninklyke Societeit in Engeland* (Leiden, 1684), Opdragt.
39 Ibid.
40 'Den drukker aan den leeser', ibid.; Douglas Anderson, 'Daniel van Gaesbeeck', *Lens on Leeuwenhoek*, https://lensonleeuwenhoek.net, accessed 23 December 2022.
41 Douglas Anderson, 'Cornelis Boutesteyn', *Lens on Leeuwenhoek*, https://lensonleeuwenhoek.net, accessed 23 December 2022.
42 Jan Albert Gruys and Johannes Joseph Maria Bos, *Adresboek Nederlandse drukkers en boekverkopers tot 1700* (The Hague, 1999), 51.
43 Douglas Anderson, 'Hendrik van Kroonevelt', *Lens on Leeuwenhoek*, https://lensonleeuwenhoek.net, accessed 23 December 2022.
44 Petra Beydals, 'Twee testamenten van Antoni van Leeuwenhoeck', *Nederlands Tijdschrift voor Geneeskunde* 77 (1933), 1027; Zacharias Konrad von Uffenbach, *Merkwürdige Reisen durch Niedersachsen, Holland und Engelland*, vol. III (Ulm, 1754), 338.
45 Antoni van Leeuwenhoek, *Send-Brieven*; Antoni van Leeuwenhoek, *Epistolae physiologicae super compluribus naturae arcanis* (Delft, 1719).
46 T. vander Wilt, 'Op de Titel-prent', in Van Leeuwenhoek, *Send-Brieven*.
47 J. G. Kerckherdere and Arnold Hoogvliet (translator), 'Lofdicht op den zeer vermaarden en in kunst en wetenschap ervaren Antoni van Leeuwenhoek', Arnold Hoogvliet, 'Dankoffer', H. K. Poot, 'Eerkroon', in *Lauwerkranssen, gevlochten Voor den heer Antoni van Leeuwenhoek, groot wysgeer, lidt der Koningklyke Gemeenschap te Londen* (Rotterdam, 1717), 13, 19, 25.
48 Ibid., 12; Arnold Hoogvliet, 'Op de brieven van de zeer geleerden heere Antoni van Leeuwenhoek', in Van Leeuwenhoek, *Send-Brieven*, 12.
49 Beydals, 'Twee testamenten van Antoni van Leeuwenhoeck', 1029.
50 Ibid.
51 Douglas Anderson, 'What happened to his papers?', *Lens on Leeuwenhoek*, https://lensonleeuwenhoek.net, accessed 25 December 2022.
52 Letter no. L-540 [XXVIII], 28 September 1716, *Alle de brieven* 18.
53 James Jurin to Maria van Leeuwenhoek, letter no. L-596 November 1723, *Alle de brieven* 19.
54 Letter no. 287, 18 August 1711, *Alle de brieven* 16.
55 Letter no. 287, 16.

9: Little Animals and Cells

1 Martin Folkes, 'Some account of Mr. Leeuwenhoek's curious microscopes, lately presented to the Royal Society', *Philosophical Transactions of the Royal Society of London* 32, 380 (31 December 1723), 446–53, in particular 447; letter no. L-591, Maria van Leeuwenhoek to the Royal Society, 4 October 1723, *Alle de brieven* 19; letter no. 228 [140], 2 August 1701, *Alle de brieven* 14.
2 Folkes, 'Some account of Mr. Leeuwenhoek's curious microscopes', 447–9.
3 Ibid., 447.
4 G. A. Lindeboom, 'De "ziekte van Van Leeuwenhoek"', *Nederlands Tijdschrift voor Geneeskunde* 119 (1975), 202–3.
5 Letter L-585, May 1723, *Alle de brieven* 19.
6 Letter L-588, August 1712, *Alle de brieven* 19.
7 Ibid.; letter no. 234 [145], 14 February 1702, *Alle de brieven* 14, 80, 82.
8 Letter no. 234 [145], 80, 82.
9 Letter no. L-588, 19.
10 Ibid.
11 Letter no. L-596, James Jurin to Maria van Leeuwenhoek, 29 November 1723, *Alle de brieven* 19.
12 Ibid.
13 Brian J. Ford, *The Leeuwenhoek Legacy* (Bristol, 1991), 51–69.
14 Marc Ratcliff, *The Quest for the Invisible: Microscopy in the Enlightenment* (Farnham, UK, and Burlington, VT, 2009), 30.
15 Ibid., 73.
16 Ibid., 168–9.
17 Letter no. 147 [90], 10 July 1695, *Alle de brieven* 10, 268, 270; Antoni spoke of more than thirty cherry trees. They would not have fitted in the small garden at his house. He also says that he cut off a sprig of a cherry tree to study when he 'came home', so he was clearly not at home when he examined the trees. From 1664 he had a garden outside the city, but he sold that in 1693, so this cannot be the garden he refers to in his letter from 1695, in which he describes research conducted in the previous summer and winter (letter no. 147 [90], 272, 276). The cherry trees and redcurrant bushes were probably in a garden that he owned at an unknown location in the city: letter no. 57 [30], 5 April 1680, *Alle de brieven* 3, 212, note 43.
18 Letter no. 147 [90], 270, 300.
19 Ibid., 270, 276, 278.
20 Letter no. 248, 21 March 1704, *Alle de brieven* 14, 334.
21 Letter no. 147 [90], 272, note 28.
22 Letter no. 105 [60], 28 November 1687, *Alle de brieven* 7, 144.
23 Letter no. 248, 334.
24 Ibid., note 72.
25 Ratcliff, *The Quest for the Invisible*, 63.
26 Ibid., 125, 126, 128.

27 Ibid., 128.
28 Ibid., 218.
29 Matthew Cobb, *Generation: The Seventeenth-Century Scientists Who Unraveled the Secrets of Sex, Life, and Growth* (New York, 2006), 242.
30 Robert Hooke, *Micrographia* (London, 1665), 113, 116.
31 William J. Croft, *Under the Microscope: A Brief History of Microscopy* (Hackensack, NJ, 2006), 8.
32 Gerard van Doornum, Neeraja Sankaran and Ton van Helvoort, *Leeuwenhoek's Legatees and Beijerinck's Beneficiaries: A History of Medical Virology in the Netherlands* (Amsterdam, 2020), 32.
33 John Edward Fletcher et al., *A Study of the Life and Works of Athanasius Kircher, 'Germanus Incredibilis': With a Selection of His Unpublished Correspondence and an Annotated Translation of His Autobiography* (Leiden, 2011), 106.
34 Athanasius Kircher, *Scrutinium Physico-Medicum Contagiosae Luis, Quae Pestis Dicitur* (Rome, 1658); Fletcher et al., *A Study of the Life and Works of Athanasius Kircher, 'Germanus Incredibilis'*, 116–18.
35 Fletcher et al., *A Study of the Life and Works of Athanasius Kircher, 'Germanus Incredibilis'*, 118.
36 Lesley Robertson et al., *Antoni van Leeuwenhoek: Master of the Minuscule* (Leiden, 2016), 80; letter no. 42 [27], 21 February 1679, *Alle de brieven* 2, 412.
37 Letter no. 42 [27], 412.
38 Ibid., 412.
39 Letter no. L-577, 13 June 1722, *Alle de brieven* 19.
40 Emanuel Timonius, 'V. An account, or history, of the procuring the small-pox by incision, or inoculation; as It has for some time been practised at Constantinople', *Philosophical Transactions of the Royal Society of London* 29, 339 (31 December 1714), 72–82.
41 Ibid., 72–4.
42 Letter no. L-575, 26 May 1722, *Alle de brieven* 19.
43 Letter no. L-577.
44 Ibid.
45 'Leeuwenhoek's jubilé', *Delftsche Courant*, 6 August 1875; 'Een feest der wetenschap', *Algemeen Handelsblad*, 9 September 1875.
46 P. J. Haaxman, *Antony van Leeuwenhoek. De ontdekker der infusoriën, 1675–1875* (Leiden, 1875).
47 Clifford Dobell, *Antony van Leeuwenhoek and His 'Little Animals': Being Some Account of the Father of Protozoology and Bacteriology and His Multifarious Discoveries in These Disciplines* (New York, 1932).
48 A. Schierbeek, *Antoni van Leeuwenhoek. Zijn leven en zijn werken* (Lochem, 1950); Lesley Robertson et al., *Van Leeuwenhoek: Groots in het kleine* (Amsterdam, 2014); Dirk van Delft, *Onzichtbaar leven: Antoni van Leeuwenhoek en de wondere wereld van de microbiologie* (Amsterdam, 2022).

49 P. Baas, 'Leeuwenhoek's contributions to wood anatomy and his ideas on sap transport in plants', in *Antoni van Leeuwenhoek 1632–1723: Studies on the Life and Work of the Delft Scientist Commemorating the 350th Anniversary of His Birthday*, ed. L. C. Palm and H.A.M. Snelders (Amsterdam, 1982), 79–107, in particular 103.
50 Antoni van Leeuwenhoek, *Send-Brieven zoo aan de hoog edele heeren van de Koninklyke societeit te Londen als aan andere aansienelyke en geleerde lieden, over verscheyde verborgentheden der natuure* (Delft, 1718).
51 Folkes, 'Some account of Mr. Leeuwenhoek's curious microscopes', 451.
52 Ibid., 452.
53 Ibid.
54 Eltjo Aldegondus van Beresteyn, *Grafmonumenten en grafzerken in de Oude Kerk te Delft* (Assen, 1938), 18.
55 Ibid., 19.

Acknowledgements

1 Gerard van Rijnberk, 'Algemeene inleiding', *Alle de brieven* 1, 16.

BIBLIOGRAPHY

Unpublished sources

Information from baptism, marriage and funeral records can be accessed at www.wiewaswie.nl/en

Municipal archives

Archief Weeskamer Delft, Delft Municipal Archives, 454 f. 373

'Begraafboeken Oude en Nieuwe Kerk Delft, mei 1628–1 januari 1644', Stadsarchief Delft, 14-38

'Begraafboeken Oude en Nieuwe Kerk Delft, 4 januari 1644–30 januari 1656', Stadsarchief Delft, 14-39

'Begraafboeken Oude en Nieuwe Kerk Delft, 1 februari 1655–30 oktober 1666', Stadsarchief Delft, 14-40

'Doopboeken Nieuwe Kerk Delft, 9 oktober 1616–31 maart 1624', Stadsarchief Delft, 14-54

'Doopboeken Nieuwe Kerk Delft, april 1624–10 augustus 1636', Stadsarchief Delft, 14-55

'Doopboeken Nieuwe Kerk Delft, 14 augustus 1636–1649', Stadsarchief Delft, 14-56

'Doopboeken Nieuwe Kerk Delft, 1661–1682', Stadsarchief Delft, 14-58

'Doopboeken Oude Kerk Delft, 1 augustus 1624–31 juli 1642', Stadsarchief Delft, 14-8

'Doopboeken Oude Kerk Delft, 1 augustus 1642–27 augustus 1658', Stadsarchief Delft, 14-9

'Doopboeken Oude Kerk Delft, 1 september 1658–18 juni 1664', Stadsarchief Delft, 14-10

'Doopboeken Oude Kerk Delft, 10 juni 1664–31 december 1684', Stadsarchief Delft, 14-11

'Kohier van de verponding op huizen en erven binnen de stad volgens het redresgeneraal van 1632 (Met Aantekeningen Tot 1656)', Stadsarchief Delft, 1-4017

'Legger van de verponding op huizen en molens te Delft, voor ontvanger Adriaen Claesz Moeijt (1620 Aangelegd, Met Aantekeningen Tot 1634)', Stadsarchief Delft, 1-4013

'Ondertrouwboek Gerecht, 10 maart 1618–26 september 1626', Stadsarchief Delft, 14-124

'Ondertrouwboek Gerecht, 1637–1645', Stadsarchief Delft, 14-001225

'Ondertrouwboek Gerecht, 1650–1656', Stadsarchief Delft, 14-27

'Ondertrouwboek Gerecht, 1670–1674', Stadsarchief Delft, 14-131

'Ondertrouwboeken Nieuwe Kerk Delft 1587–1626', Stadsarchief Delft, 14-6

'Ondertrouwboeken Nieuwe Kerk Delft 1626–1811', Stadsarchief Delft, 14-68, 69, 70

'Ondertrouwboeken Oude Kerk Delft', Stadsarchief Delft, 14-22

'Plan van David Coornwinder voor de droogmaking van Het Oostmeer in de polder van Berkel', 1641, Rotterdam Municipal Archives, 1305-388

Bogaert, Andries, 'Minuutakten notaris Andries Bogaert 1654–1658', Delft Municipal Archives, 161-1888 (70v).

Geesteranus, Joris, 'Inventaris Nalatenschap Maria van Leeuwenhoek', 26 July 1745. Old Notarial Archives Delft, 1574–1842. Stadsarchief Delft, 161-2791

Graeff, Guillaume de, 'Testament Thonis Philipsz Leeuwenhoek Sr. Bij Notaris Guillaume de Graeff', 2 April 1643, Old Notarial Archives, 161–71, Stadsarchief Delft

Ranck, Johannes, 'Inventaris van Den Boedel En Goederen van Elisabeth Cornelisdr Door Notaris Johannes Ranck', 1664, Old Notarial Archives Delft, Minuten van allerlei akten, Stadsarchief Delft, 161-2119

Trioen, Jan et al., Nodige Verantwoordinge Voor de Heer Pieter Rabus En Juffr. Syne Huysvrouw Tegens de Amsterdammers En Haarlemmers, Niet Gelovende de Werking Der Wichelroeden, 4, 7

Published letters

Antoni van Leeuwenhoek

Antoni van Leeuwenhoek and the Leeuwenhoek Committee, *Alle de brieven van Antoni van Leeuwenhoek* vols I–XX (1939–2023)

Leeuwenhoek, Antoni van, *Ondervindingen en beschouwingen der onsigbare geschapene waarheden, waar in gehandelt wert vande schobbens inde mond, de lasarie, de jeuking, 't kind met vis-schobbens, 't binnenste der darmen, en de beweging derselve, als mede het vet dat inde selve gevonden wert: Geschreven aande wyt-beroemde Koninklyke Societeit in Engeland* (Leiden, 1684)

Leeuwenhoek, Antoni van, *Arcana Naturae Detecta Ab Antonio van Leeuwenhoek* (Delft, 1695)

Leeuwenhoek, Antoni van, *Send-Brieven zoo aan de Hoog Edele heeren van de Koninklyke Societeit te Londen als aan andere aansienelyke en geleerde lieden, over verscheyde verborgentheden der natuure* (Delft, 1718)

Leeuwenhoek, Antoni van, *Epistolae Physiologicae Super Compluribus Naturae Arcanis* (Delft, 1719)

Other

Huygens, Christiaan, *Oeuvres Complètes*, vol. I: *Correspondance, 1638–1656* (The Hague, 1888)
——, *Oeuvres complètes*, vol. III: *Correspondance, 1676–84* (The Hague, 1899)
——, *Oeuvres Complètes*, vol. VIII: *Correspondance, 1676–1684* (The Hague, 1899)
Leibniz, Gottfried Wilhelm, *Sämtliche Schriften und Briefe. Dritte Reihe. Mathematischer, naturwissenschaftlicher und technischer Briefwechsel*, vol. V: *1691–3* (Berlin, 2003)
Swammerdam, Jan, Melchisédech Thévenot and Gerrit Arie Lindeboom, *The Letters of Jan Swammerdam to Melchisedec Thévenot: With English Translation and a Biographical Sketch* (Amsterdam, 1975)

Primary literature

'Bekentmaking', *Hollandsche Historische Courant*, 16 May 1747
Catalogus van het vermaarde cabinet van vergrootglasen, met zeer veel moeite, en kosten in veele jaren geïnventeert, gemaakt, en nagelaten door wylen den heer Anthony van Leeuwenhoek, in zyn Ed: Leeven Lid van de Koninglyke Societeit der Wetenschappen in Londen (Delft, 1747)
'Een feest der wetenschap', *Algemeen Handelsblad*, 9 September 1875
Lauwerkranssen, gevlochten voor den heer Antoni van Leeuwenhoek, groot wysgeer, lidt der Koningklyke Gemeenschap te Londen (Rotterdam, 1717)
'Leeuwenhoek's Jubilé', *Delftsche Courant*, 4 August 1875
Beeckman, Isaac, and Cornelis de Waard, eds, *Journal tenu par Isaac Beeckman de 1604 à 1634*, vol. III (The Hague, 1945)
Bekker, Balthasar, '4, Waar in 't bewijs, dat uit d'Ervarentheid genomen word, ten gronde toe word ondersocht', in *De betoverde wereld* (Amsterdam, 1691)
Bontekoe, Cornelis, *Reden over de koortzen; Door welke aangewesen word, dat de gemene theorie en praktijk valsch, schadelijk en moordadig is* (The Hague, 1682)
——, *Tractaat van het excellentste kruyd thee* (The Hague, 1678)
Boyle, Robert, 'Shewing the occasion of making this new essay – instrument, together with the hydrostatical principle 'tis founded on', *Philosophical Transactions* 10, 115 (1675), 329–48
Folkes, Martin, 'Some account of Mr. Leeuwenhoek's curious microscopes, lately presented to the Royal Society', *Philosophical Transactions of the Royal Society of London* 32, 380 (31 December 1723), 446–53
Graaf, Regnier de, *Alle de wercken, so in de ontleed-kunde, als andere deelen der medicyne* (Amsterdam, 1686)
——, *Tractatus anatomico-medicus de succi pancreatici natura et usu* (Leiden, 1671)
Hartsoeker, Nicolaas, *Eclairissement sur les conjectures physiques* (Amsterdam, 1710)
——, *Essay de dioptrique* (Paris, 1694)

—, 'Extrait critique des lettres de feu M. Leeuwenhoek', in *Cours de physique accompagné de plusieurs piéces concernant la physique qui ont déja paru et d'un extrait critique des lettres de M. Leeuwenhoek* (The Hague, 1730)

Hartsoeker, Nicolaas, and Alexander Block (translation), *Proeve der deurzicht-Kunde* (Amsterdam, 1699)

Hooke, Robert, *Micrographia: or Some Physiological Descriptions of Minute Bodies Made By Magnifying Glasses With Observations and Enquiries thereupon* (London, 1665)

Huygens, Christiaan, 'Extrait d'une Lettre de M. Huguens de l'Acad. R. des Sciences à l'Auteur du Journal, touchant une nouvelle maniere de Microscope qu'il a apporté de Hollande', *Journal des sçavans* (15 August 1678), 345–7

Kircher, Athanasius, *Scrutinium Physico-Medicum Contagiosae Luis, Quae Pestis Dicitur* (Rome, 1658)

Leeuwenhoek, Antoni van, 'Observations, communicated to the publisher by Mr. Antoni van Leewenhoeck, in a Dutch Letter of the 9th Octob. 1676. Here English'd: concerning little Animals by Him Observed in rain-well-sea-and-snow water; as also in water wherein pepper had lain infused', *Philosophical Transactions* 12, 133 (25 March 1677), 821–31

Loon, Gerard van, *Beschryving der Nederlandsche historiepenningen. Of beknopt verhaal van 't gene sedert de overdracht der heerschappye van keyzer Karel den Vyfden op koning Philips zynen zoon, tot het sluyten van den Uytrechtschen vreede, in de zeventien Nederlandsche gewesten is voorgevallen*, vol. IV (The Hague, 1731)

Nispen, Mattheus van, *De beknopte lant-meet-konst: Leerende in t' korte, alles wat in t' gemeen, in de practijcke des landt-metens voor-komen kan* (Dordrecht, 1662)

Rabus, Pieter, *Boekzaal van Europe*, January–February 1697

—, *Boekzaal van Europe*, May–June 1697

—, *Boekzaal van Europe*, October 1697

Spoors, Jacob, *'Caart ende ontworp gedaan bij mijn Jacob Spoors geadmitteert landtmeter bijden Hove van Hollandt tot Delft residerende vanhet noordteinde van Berckel met affteickeninge van een geconcipieerde nieuwe te make vaart ofte watering, vande Berckelse kade aff in het noorde ende soo voorts zuijdwaarts op tot inde Oostermeer van Berckel, sooals die op de kaart affgeteickent is, doch alles op approbatie vande wel edele hoge heemraden van Delfflandt'* (1675)

—, *Oratie van de nieuwe wonderen des wereldts, de nuttigheyd, de waerdigheyd, der wis- ende meetkunsten* (Delft, 1638)

Swammerdam, Jan, *Bybel der natuure door Jan Swammerdam, Amsteldammer, of historie der insecten*, vol. II (Leiden, 1737)

Timonius, Emanuel, 'V. An account, or history, of the procuring the smallpox by incision, or inoculation; as It has for some time been practised at Constantinople', *Philosophical Transactions of the Royal Society of London* 29, 339 (31 December 1714), 72–82

Uffenbach, Zacharias Konrad von, *Merkwürdige Reisen durch Niedersachsen, Holland und Engelland*, vol. III (Ulm, 1754)

Websites

Douglas Anderson, Lens on Leeuwenhoek

'Appointed co-wine gauger to help Dirk Arisz', https://lensonleeuwenhoek.net, accessed 19 December 2022

'Leeuwenhoek's Cabinet of Wonders', https://lensonleeuwenhoek.net/content, accessed 6 December 2022

'Cornelis Boutesteyn', https://lensonleeuwenhoek.net, accessed 23 December 2022

'Daniel van Gaesbeeck', https://lensonleeuwenhoek.net, accessed 23 December 2022

'Hendrik van Kroonevelt', https://lensonleeuwenhoek.net, accessed 23 December 2022

'Hippolytusbuurt 3, Leeuwenhoek's Home and Laboratory', https://lensonleeuwenhoek.net, accessed 14 December 2022

'Inherited property on Oosteinde from first wife Barbara de Meij's family', https://lensonleeuwenhoek.net, accessed 14 December 2022

'Jacob Spoors', https://lensonleeuwenhoek.net, accessed 22 December 2022

'Jan Jacobs de Molijn', https://lensonleeuwenhoek.net, accessed 14 December 2022

'St. Nicolaas Gilde', https://lensonleeuwenhoek.net, accessed 19 October 2022

'Stepfather Jacob Jans de Molijn appeared before Weeskamer to appoint children's guardians', https://lensonleeuwenhoek.net, accessed 14 December 2022

'Thonis Philips Leeuwenhoek', https://lensonleeuwenhoek.net, accessed 13 December 2022

'What happened to his papers?', https://lensonleeuwenhoek.net, accessed 25 December 2022

Other

GUM, 'Deelcollectie Geschiedenis van de Wetenschappen', www.gum.gent, accessed 5 December 2022

Leerintveld, Ad, 'Huygens en Van Leeuwenhoek', Brieven Constantijn Huygens Online, https://brievenconstantijnhuygens.net, accessed 22 July 2022

Visualizing the Unknown, 'MicroLab3: Bodily Fluids', https://visualizingthe-unknown.com, accessed 9 November 2022

WieWasWie, https://wiewaswie.nl, accessed 9 March 2021

Secondary literature

Abou-Nemeh, S. Catherine, 'The natural philosopher and the microscope: Nicolas Hartsoeker unravels nature's "Admirable OEconomy"', *History of Science* 51, 1 (2013), 1–32

Aillaud, Gilles et al., *Vermeer* (Paris, 1987)
Aldersey-Williams, Hugh, *Dutch Light: Christiaan Huygens and the Making of Science in Europe* (London, 2020)
Anderson, Douglas, 'The early civic career of Anthony van Leeuwenhoek: The case of Sijmon Bourbon [1667–70]', article under preparation
—, 'Still going strong: Leeuwenhoek at eighty', *Antonie van Leeuwenhoek. Journal of Microbiology* 106 (2014), 3–26
—, 'The tensions between facts and fantasy', *Studium* 8, 4 (2016), 227–30
—, '"Your Most Humble Servant": The letters of Antoni van Leeuwenhoek', *FEMS Microbiology Letters* 369, 1 (2022)
Bakker, Nelleke, Jan Noordman and Marjoke Rietveld-Van Wingerden, *Vijf eeuwen opvoeden in Nederland: Idee en praktijk 1500–2000* (Assen, 2010)
Bartjens, Willem, Danny Beckers and Marjolein Kool, *De cijfferinghe (1604). Het rekenboek van de beroemde schoolmeester Willem Bartjens. Ingeleid door Danny Beckers en Marjolein Kool* (Hilversum, 2004)
Bedini, Silvio A., 'Galileo and scientific instrumentation', in *Reinterpreting Galileo, Studies in Philosophy and the History of Philosophy* 15 , ed. William A. Wallace (Washington, DC, 1986), 127–53
Bekkering, Harry et al., eds, *De hele Bibelebontse berg. De geschiedenis van het kinderboek in Nederland & Vlaanderen van de middeleeuwen tot heden* (Amsterdam, 1989)
Beresteyn, Eltjo Aldegondus van, *Grafmonumenten en grafzerken in de Oude Kerk te Delft* (Assen, 1938)
Bergvelt, Ellinoor, and René Kistemaker, eds, *De wereld binnen handbereik: Nederlandse kunst- en rariteitenverzamelingen, 1585–1735* (Zwolle and Amsterdam, 1992)
Bergvelt, Ellinoor, Debora J. Meijers and Alexis Joachimides, eds, *Kabinetten, galerijen en musea: Het verzamelen en presenteren van naturalia en kunst van 1500 tot heden*, revd edn (Zwolle, 2013)
Berkel, Klaas van, *Citaten uit het boek der natuur: opstellen over Nederlandse wetenschapsgeschiedenis* (Amsterdam, 1998)
Berlinghoff, William P., Fernando Q. Gouvêa and Desiree van den Bogaart, *Wortels van de wiskunde: Een historisch overzicht voor leraren en anderen* (Amsterdam, 2016)
Beydals, Petra, 'Twee testamenten van Antoni van Leeuwenhoeck', *Nederlands Tijdschrift voor Geneeskunde* 77 (1933), 1021–33
Birch, Thomas, *The History of the Royal Society of London*, vol. III (London, 1757)
—, *The History of the Royal Society of London*, vol. IV (London, 1757)
Boitet, Reinier, *Beschryving der stadt Delft, behelzende een zeer naaukeurige en uitvoerige verhandeling van deszelfs eerste oorsprong, benaming, bevolking, aanwas, gelegenheid, prachtige en kunstige gedenkstukken en zeltzaamheden. Nevens derzelver voorregten, handvesten, previlegien, en regeeringsvorm: alles 't zamengestelt en getrokken uit oude handtschriften,*

memorien, en brieven, en met zeer veele echte bewysstukken (te vooren noit gedrukt) bevestigt: door verscheide liefhebbers en kenners der Nederlandsche oudheden (Delft, 1729)

Booy, Engelina Petronella de, *Kweekhoven der wijsheid. Basis- en vervolgonderwijs in de steden van de provincie Utrecht van 1580 tot het begin der 19e eeuw* (Zutphen, 1980)

Bots, Hans, *De Republiek der Letteren: De Europese intellectuele wereld 1500–1760* (Nijmegen, 2018)

Braeckman, Johan, and Linda Van Speybroeck, eds, *Fascinerend leven: Een geschiedenis van de biologie* (Gent, 2021)

Cobb, Matthew, *Generation: The Seventeenth-Century Scientists Who Unraveled the Secrets of Sex, Life, and Growth*, 1st U.S. edn (New York, 2006)

Cocquyt, Tiemen, 'De identificatie van een zilveren microscoopje van Antoni van Leeuwenhoek (1632–1723)', *Studium* 8, 4 (2016), 198–211

—, 'Positioning Van Leeuwenhoek's Microscopes in 17th-century microscopic practice', *FEMS Microbiology Letters* 369, 1 (2022)

Cocquyt, Tiemen, Marvin Bolt and Michael Korey, 'Hudde en zijn gesmolten microscooplensjes', *Studium* 11, 1 (2018), 78–95

Cocquyt, Tiemen et al., 'Neutron tomography of Van Leeuwenhoek's microscopes', *Science Advances* 7, 20 (2021)

Cole, Francis, *Early Theories of Sexual Generation* (Oxford, 1930)

Croft, William J., *Under the Microscope: A Brief History of Microscopy* (Hackensack, NJ, 2006)

De Vries, Jan, and A. M. van der Woude, *Nederland 1500–1815: De eerste ronde van moderne economische groei*, 3rd edn (Amsterdam, 2005)

Delft, Dirk van, *Onzichtbaar leven: Antoni van Leeuwenhoek en de wondere wereld van de microbiologie* (Amsterdam, 2022)

Deursen, A. Th. van, 'Mensen van klein vermogen. Het kopergeld van de Gouden Eeuw', in *De Gouden Eeuw Compleet* (Amsterdam, 2010), 7–459

Dobell, Clifford, *Antony van Leeuwenhoek and His 'Little Animals', Being Some Account of the Father of Protozoology and Bacteriology and His Multifarious Discoveries in These Disciplines*, new edn (New York, 1960)

Doornum, Gerard van, Neeraja Sankaran and Ton van Helvoort, *Leeuwenhoek's Legatees and Beijerinck's Beneficiaries: A History of Medical Virology in the Netherlands* (Amsterdam, 2020)

Egmond, Wim van, 'The riddle of the "Green Streaks"', *Micscape*, 239 (2016)

Elmer, Peter, ed., *The Healing Arts: Health, Disease and Society in Europe 1500–1800* (Manchester, 2004)

Fletcher, John Edward, et al., *A Study of the Life and Works of Athanasius Kircher, 'Germanus Incredibilis': With a Selection of His Unpublished Correspondence and an Annotated Translation of His Autobiography* (Leiden, 2011)

Ford, Brian J., 'Antoni van Leeuwenhoek – microscopist and visionary scientist', *Journal of Biological Education* 23, 4 (1989), 293–9

—, *The Leeuwenhoek Legacy* (Bristol, 1991)

Fournier, Marian, *The Fabric of Life: The Rise and Decline of Seventeenth-Century Microscopy* (Enschede, 1991)

Fransen, Sietske, 'Antoni van Leeuwenhoek, his images and Draughtsmen', *Perspectives on Science* 27, 3 (2019), 485–544

Galluzzi, Paolo, and Peter Mason, *The Lynx and the Telescope: The Parallel Worlds of Federico Cesi and Galileo* (Leiden and Boston, MA, 2017)

Goldgar, Anne, *Impolite Learning: Conduct and Community in the Republic of Letters, 1680–1750* (New Haven, CT, 1995)

Gruys, Jan Albert, and Johannes Joseph Maria Bos, *Adresboek Nederlandse drukkers en boekverkopers tot 1700* (The Hague, 1999)

Haaxman, P. J., *Antoni van Leeuwenhoek. De ontdekker der infusoriën, 1675–1875* (Leiden, 1875)

Halbertsma, H., 'Ontleedkundige aanteekeningen. VI. Johan Ham van Arnhem, de ontdekker der spermatozoïden', *Verslagen en mededeelingen der Koninklijke Akademie van Wetenschappen. Afdeeling Natuurkunde* XIII (1862)

Heniger, J., 'Antoni van Leeuwenhoek en zijn diploma van de Royal Society', in *Tijdschrift voor de geschiedenis der geneeskunde, natuurwetenschappen, wiskunde en techniek*, 1 (1978), 157–69

Hillen, H.F.P., E. S. Houwaart and F. G. Huisman, *Medische geschiedenis: Ziekte, kennis, dokter en patiënt, gezondheidszorg en maatschappij* (Houten, 2018)

Houtzager, Hendrik Leonard, ed., *Delft en de Oostindische Compagnie* (Amsterdam, 1987)

—, *Reinier de Graaf, 1641–1673: in sijn leven nauwkeurig ontleder en gelukkig geneesheer tot Delft; bundel opstellen, verschenen bij gelegenheid van de herdenking van de 350ste geboortedag van Reinier de Graaf* (Rotterdam, 1991)

Hunter, Michael, and Edward B. Davis, eds, *The Works of Robert Boyle: Electronic Edition*, vol. 1, Publications to 1660 (London, 1999)

Huygens, Constantijn, and Chris L. Heesakkers, *Mijn jeugd* (Amsterdam, 1987)

Jorink, Eric, 'Swammerdam, hoveling? Enige kanttekeningen bij de reputatie van een wetenschappelijk onderzoeker', *Studium* 8, 4 (2016), 173–97

—, '"These wonderful glasses": Dutch humanists and the microscope, 1620–1670', in *Who Needs Scientific Instruments? Conference on Scientific Instruments and Their Users, 20–22 October 2005*, ed. Bart Grob and Hans Hooijmaijers (Leiden, 2006)

King, Peter, *The Life of John Locke, With Extracts From His Correspondence, Journals, and Common-Place Books*, vol. 1 (London, 1830)

Konings, Gertrud, 'Opuntia Ficus-Indica as an extremophile: A survivor in extreme climates?', *Cactus and Succulent Journal* 82, 4 (2010), 172–5

Kooijmans, Luuc, *Gevaarlijke kennis: Inzicht en angst in de dagen van Jan Swammerdam* (Amsterdam, 2007)

Leong, Elaine Yuen Tien, and Alisha Michelle Rankin, eds, *Secrets and Knowledge in Medicine and Science, 1500–1800* (Farnham, Surrey, UK, and Burlington, VT, 2011)

Lijn, P. van der, *Nederlandse zwerfstenen* (Zutphen, 1935)

Lindeboom, G. A., 'Reinier de Graaf (1641–1673)', *Nederlands Tijdschrift voor Geneeskunde* 118, 21 (1974), 789–95

——, 'De "ziekte van Van Leeuwenhoek"', *Nederlands Tijdschrift voor Geneeskunde* 119 (1975), 202–3

Lindemann, Mary, *Medicine and Society in Early Modern Europe* (Cambridge, 2020)

Lintsen, Harry, ed., *Geschiedenis van de techniek in Nederland 5: Techniek, beroep en praktijk* (The Hague, 1994)

Meurer, Peter H., 'Cornelis van Beughem, een vergeten randfiguur van de Amsterdamse kartografie', *Caert Thresoor. Tijdschrift voor de Geschiedenis van de Kartografie* 31, 2 (2012), 39–46

Stedelijk Museum Het Prinsenhof, *De stad Delft. Cultuur en maatschappij van 1572 tot 1667* , ed. I. Spaander (Delft, 1981)

Neogi, Tuhina et al., 'Alcohol quantity and type on risk of recurrent gout attacks: An internet-based case-crossover study', *American Journal of Medicine* 127, 4 (2014), 311–18

Newton, Hannah, *The Sick Child in Early Modern England, 1580–1720* (Oxford, 2012)

Noordegraaf, Leo, and Gerrit Valk, *De gave Gods: de pest in Holland vanaf de late middeleeuwen*, 2nd revd edn (Amsterdam, 1996)

Obreen, D. O., 'Het Sint Lucas-Gild te Delft', in *Archief Voor Nederlandsche kunstgeschiedenis* 1 (Rotterdam, 1877), 1–120

Palm, Lodewijk C., 'Antoni van Leeuwenhoeks kristallographische und mineralogische Untersuchungen', *Montanmedizin und Bergbauwissenschaften: Hallesches Symposium* (1987), 111–19

Palm, Lodewijk C., and H.A.M. Snelders, eds, *Antoni van Leeuwenhoek 1632–1723: Studies on the Life and Work of the Delft Scientist Commemorating the 350th Anniversary of His Birthday* (Amsterdam, 1982)

Parke, Emily C., 'Flies from meat and wasps from trees: Reevaluating Francesco Redi's spontaneous generation experiments', *Studies in History and Philosophy of Science Part C: Studies in History and Philosophy of Biological and Biomedical Sciences* 45 (2014), 34–42

Philippa, Marlies, Frans Debrabandere and Arend Quak, *Etymologisch woordenboek van het Nederlands* (Amsterdam, 2005)

Polanen, Tim van, 'Snak, Claas and Bastiaan's struggle for freedom: Three Curaçaoan enslaved men and their court cases about the free soil principle in the Dutch Republic', *BMGN – Low Countries Historical Review* 136, 1 (2021), 33–58

Principe, Lawrence, ed., *Chymists and Chymistry: Studies in the History of Alchemy and Early Modern Chemistry* (Sagamore Beach, MA, 2007)

Ragland, Evan, 'Experimenting with chymical bodies: Reinier de Graaf's investigations of the pancreas', *Early Science and Medicine* 13, 6 (2008), 615–64

Ratcliff, Marc, *The Quest for the Invisible: Microscopy in the Enlightenment* (Farnham, UK, and Burlington, VT, 2009)

Reith, Reinhold, 'Circulation of skilled labour in late medieval and early modern Central Europe', in *Guilds, Innovation and the European Economy, 1400–1800*, ed. R. Epstein and Maarten Prak (Cambridge, 2008), 114–42

Robertson, Lesley et al., *Van Leeuwenhoek: Groots in het kleine*, Wetenschappelijke Biografie 53 (Amsterdam, 2014)

—, *Antoni van Leeuwenhoek: Master of the Minuscule* (Leiden, 2016)

Ruestow, Edward G., *The Microscope in the Dutch Republic: The Shaping of Discovery* (Cambridge and New York, 1996)

Schierbeek, A., *Antoni van Leeuwenhoek. Zijn leven en zijn werken* (Lochem, 1950)

—, 'Een paar nieuwe bijzonderheden over Van Leeuwenhoek', *Nederlands Tijdschrift voor Geneeskunde* 74 (1930), 3891–9

Seters, W. H. van, 'Antoni van Leeuwenhoek in Amsterdam', *Notes and Records of the Royal Society of London* 9, 1 (31 October 1951), 36–45

—, 'Leeuwenhoek's afkomst en jeugd', *Biologisch Jaarboek* 19 (1952), 123–84

—, 'Van Leeuwenhoeks tweede huwelijk', *Nederlands Tijdschrift voor Geneeskunde* 112, 27 (1968), 1258–61

Shapin, Steven, *A Social History of Truth: Civility and Science in Seventeenth-Century England* (Chicago, IL, 1994)

Shapin, Steven, and Simon Schaffer, *Leviathan and the Air-Pump: Hobbes, Boyle, and the Experimental Life* (Princeton, NJ, 1989)

Shapiro, Barbara J., 'Testimony in seventeenth-century English natural philosophy: Legal origins and early development', *Studies in History and Philosophy of Science* 33 (2002), 243–63

Snelders, H.A.M., 'Christiaan Huygens and Newton's theory of gravitation', *Notes and Records of the Royal Society of London* 43, 2 (1989), 209–22

Strickland, Lloyd, Erik Vynckier and Julia Weckend, eds, *Tercentenary Essays on the Philosophy and Science of Leibniz* (New York, 2016)

Verhoeven, Gerrit, *De derde stad van Holland: Geschiedenis van Delft tot 1795* (Zwolle, 2015)

Wijsenbeek-Olthuis, Thera, *Achter de gevels van Delft: Bezit en bestaan van rijk en arm in een periode van achteruitgang (1700–1800)* (Hilversum, 1987)

Willach, Rolf, 'The long road to the invention of the telescope', in *The Origins of the Telescope*, ed. Albert van Helden et al. (Amsterdam, 2010), 93–114

Zuidervaart, Huib J., and Douglas Anderson, 'Antony van Leeuwenhoek's microscopes and other scientific instruments: New information from the Delft archives', *Annals of Science* 73, 3 (2016), 257–88

Zuidervaart, Huib J., and Marlise Rijks, '"Most rare workmen": Optical practitioners in early seventeenth-century Delft', *British Journal for the History of Science* 48, 1 (2015), 53–85

Zwier, Karen R., 'Methodology in Aristotle's theory of spontaneous generation', *Journal of the History of Biology* 51, 2 (2018), 355–86

ACKNOWLEDGEMENTS

A little less than a century ago, admirers of Antoni van Leeuwenhoek decided to set up a 'memorial' to honour him, 'a work unique in its kind.'[1] That memorial was to become *Alle de brieven van Antoni van Leeuwenhoek*, a twenty-part series with hundreds of letters and extensive explanatory notes. Their work, and that of their successors, made mine very much easier. I would therefore like first to thank all those who contributed to the series.

The last parts of the series are the work of Douglas Anderson and Huib Zuidervaart. Without them, the letter project would never have been completed, and this book would not have been written. The plan to write it was their idea, and they also did a great deal of research, which I was very grateful to make use of. Douglas's site *Lens on Leeuwenhoek* (www.lensonleeuwenhoek.net) in particular was a gold mine. By far not all the information that he has gathered together found its way into this book, so anyone wishing to know more can best start there.

Douglas and Huib also read my text, together with Lodewijk Palm, who was also responsible for parts of *Alle de brieven*. I could not have wished for better informed and constructive critics, and I am very grateful for their improvements and additions.

Then there was the team from Visualizing the Unknown, the research project on microscopy in the seventeenth century. The 'MicroLabs' organized by the project brought Antoni's work to life for me. It was a privilege to watch experts who were many times more proficient than I was in preparing specimens and working with microscopes, and to exchange ideas with historical specialists. So my heartfelt thanks to Eric Jorink, Sietske Fransen, Tiemen Cocquyt, Mieneke ten Hennepe, Tim Huisman, Wim van Egmond, Frank van Campen, Hans Huijbregts, Ellen Pater, Erma Hermens and Larissa van Vianen.

I would also like to thank the Rijksmuseum Boerhaave for its support, including with the launch of this book.

I am very grateful to the Prince Bernhard Culture Fund and all its donors for their financial support, which enabled me to study texts by and about Antoni in great depth.

The enthusiasm of Alexandra van Dam Merrett and Léon Groen at Het Spectrum publishing house was infectious, and I am delighted at the care they took in preparing the Dutch version of the book. My heartfelt thanks, too, to Erica van Rijsewijk for her meticulous correction of the text.

Casper Heimel impressed me immensely with his Python skills and his help in analysing the letters. Thank you, Casper! And finally, to my always encouraging readers Wil Dekkers and Elly Roose: the greatest of thanks.

LIST OF ILLUSTRATIONS

Illustration 12: From Francesco Stelluti, *Melissographia* (Rome, 1625). Biblioteca Apostolica Vaticana, Vatican City, photo © NPL – De Agostini Picture Library/Bridgeman Images.

Illustration 13: Museum Boijmans Van Beuningen, Rotterdam (from the estate of F.J.O. Boijmans, NV 5 (PK)), photo Studio Tromp.

Illustration 14: Drawing in letter no. 16 [10], 11 February 1675. From *Alle de brieven van Antoni van Leeuwenhoek* (*The Collected Letters of Antoni van Leeuwenhoek*), vol. I (1939), plate XXV, photo Wellcome Library, London.

Illustration 15: From Antoni van Leeuwenhoek, *Continuatio arcanorum naturae detectorum* (Delft, 1697), between pp. 124 and 125, photo Zentralbibliothek Zurich.

Illustration 16: Ink drawing in the margin of letter no. 3, 15 August 1673. From *Alle de brieven van Antoni van Leeuwenhoek*, vol. I (1939), plate V, photo Wellcome Library, London.

Illustration 17: From *Philosophical Transactions of the Royal Society of London*, X/115 (21 June 1675), photo Library and Archives, Natural History Museum, London.

Illustration 18: From Robert Hooke, *Micrographia* (London, 1665), between pp. 124 and 125, photo Wellcome Collection, London.

Illustration 19: Coenraet Decker, *View of the Fish Market and Meat Hall in Delft*, 1678–1703, etching. Rijksmuseum, Amsterdam (RP-P-AO-11-20).

Illustrations 20 and 21: Drawings in letter no. 214 [128], 9 July 1700. From Antoni van Leeuwenhoek, *Sevende vervolg der brieven* (Delft, 1702), between pp. 230 and 231, photo Library of Congress, Rare Book and Special Collections Division, Washington, DC.

Illustration 22: From Sebastian Münster, *Rudimenta mathematica* (Basel, 1551), p. 6, photo Sächsische Landesbibliothek – Staats- und Universitätsbibliothek, Dresden.

Illustration 23: From *Philosophical Transactions of the Royal Society of London*, XII/142 (February 1679), photo Wellcome Collection, London.

Illustration 24: Archives of Pearson Scott Foresman, donated to Wikimedia Foundation (public domain).

Illustration 37: From Antoni van Leeuwenhoek, *Arcana naturae detecta* (Delft, 1695), between pp. 192 and 193, photo Library of Congress, Rare Book and Special Collections Division, Washington, DC.

Illustration 38: From Antoni van Leeuwenhoek, *Send-Brieven, zoo aan de hoog edele Heeren van de Koninklyke Societeit te Londen* (Delft, 1718), photo Countway Library of Medicine, Harvard University, Boston, MA.

Illustration 39: Drawing in letter no. 234 [145], 14 February 1702. From Antoni van Leeuwenhoek, *Sevende vervolg der brieven* (Delft, 1702), between pp. 418 and 419, photo Library of Congress, Rare Book and Special Collections Division, Washington, DC.

Illustration 40: Johannes Jelgerhuis, *Tomb of Antoni van Leeuwenhoek*, 1820–33, lithograph. Rijksmuseum Amsterdam (RP-P-1905-2444).

Image p. 6: 'Edge of a mosquito's wing' in letter no. 70 [37], 22 January 1683. From Antoni van Leeuwenhoek, *Arcana naturae detecta* (Delft, 1695), p. 37, photo Library of Congress, Rare Book and Special Collections Division, Washington, DC.

Images p. 12: *above*: 'Spider's eyes' in letter no. 226 [138], 21 June 1701. From Antoni van Leeuwenhoek, *Sevende vervolg der brieven* (Delft, 1702), between pp. 332 and 333, University of Maryland, Baltimore Digital Archive.
left: 'Eyes of a beetle' in letter no. 193 [111], 9 May 1698. From Antoni van Leeuwenhoek, *Sevende vervolg der brieven* (Delft, 1702) between pp. 46 and 47, photo University of Maryland, Baltimore Digital Archive.
below: 'Head of an apple blossom beetle' in letter no. 144 [89], 18 May 1695. From Antoni van Leeuwenhoek, *Arcana naturae detecta* (Delft, 1695), between pp. 528 and 529, photo Library of Congress, Rare Book and Special Collections Division, Washington, DC.

Images p. 32: *above*: 'Mating dragonflies' in letter no. 65 [33], 12 November 1680. From Antoni van Leeuwenhoek, *Arcana naturae detecta* (Delft, 1695), between pp. 18 and 19, photo Library of Congress, Rare Book and Special Collections Division, Washington, DC.
below left: 'Ants' in letter no. 103 [58], 9 September 1687. From Antoni van Leeuwenhoek, *Vervolg der brieven* (Leiden, 1704), between pp. 98 and 99, photo Getty Research Institute, Los Angeles.
below right: 'Aphid' in letter no. 219 [134], 26 October 1700. From Antoni van Leeuwenhoek, *Sevende vervolg der brieven* (Delft, 1702), between pp. 286 and 287, photo University of Maryland, Baltimore Digital Archive.

Image p. 240: Detail from Jacob Hoefnagel, after Joris Hoefnagel, *Archetypa studiaque patris*, [part II, plate 4], 1592, engraving. National Gallery of Art, Washington, DC.

INDEX

Illustration numbers are indicated by *italics*